GRAPHICS & GEOMETRY 2

ENGINEERING DESIGN
GRAPHICS DEPARTMENT

TEXAS A&M UNIVERSITY

JAMES H. EARLE
J. TIMOTHY COPPINGER
PAUL M. MASON
LAWRENCE E. STARK
GEORGE J. RACZKOWSKI
JOHN T. DEMEL
THOMAS J. KRUEGER
RONALD E. BARR
ROBERT A. WILKE

Published and copyrighted by
Creative Publishing Company, 1976
ISBN 0-932702-60-0

Third printing , January 1980 (5.0xx2.70)
Second printing, September 1977 (5AB 1.74)
First printing, April 1976 (5AB2.00)

Creative Publishing Company
Box 9292, Ph 713-846-7907
College Station, Texas 77840

To the student

This problem book was designed to assist the student in thinking and communicating with the medium of graphics. In addition to covering the process of preparing working drawings, this book emphasizes descriptive geometry and the principles of analytical, plane, and solid geometry. These are essential problem-solving disciplines that must be used in the development of solutions to creative design problems.

COURSE ORGANIZATION

Your instructor will assign problems to be solved in class as daily assignments. These should be solved in accordance with his instructions. Your grade in the course will be determined as follows:

No.		
________	Daily grades	________
________	Scheduled tests	________
________	Pop tests	________
________	Design project	________
________	Optional assignments	________
		100%

Actual examples have been taken from industry to make the problems in this book as realistic as possible. All problems are printed on opaque paper; however, it is suggested that certain sheets be traced on tracing paper sheets from the back of the book and then solved in the conventional manner. These tracings can then be duplicated by available reproduction processes to demonstrate the need for good draftsmanship. Most problems can be solved with either the metric or English system of scales.

When your graded papers have been returned, you should maintain a listing of the problems and the grades earned on each on the following grade sheet. You should record the time required to solve each problem under the heading of "minutes" in the title strip in order to become familiar with the time required to solve the problems.

PROBLEM LAYOUT

Small x-marks are given on each problem sheet to show where the reference lines should be drawn to insure that the solutions will fit on each sheet. X1 is for a primary auxiliary line; X2 for a secondary; and X3 for a third successive line. The example at the right shows how these lines are used.

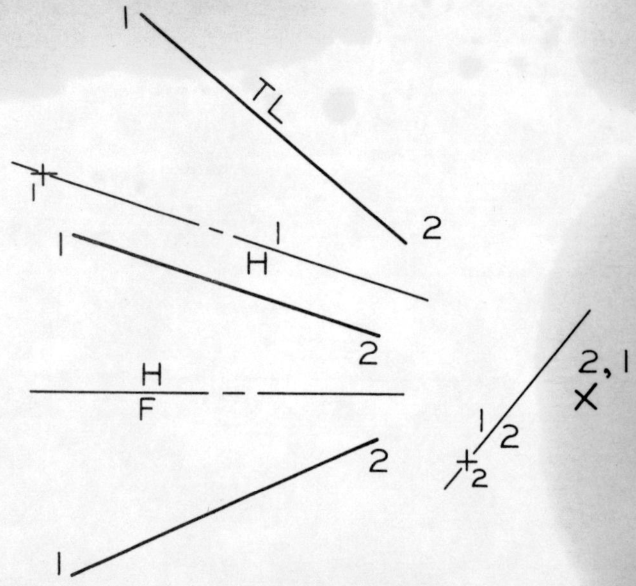

Contents

REFERENCES: The numbers at the headings of the columns above indicate the books and their respective chapters that can be used as reference in solving the various problems within this book.

1. Earle, James H., ENGINEERING DESIGN GRAPHICS (3rd Edition), Reading, Massachusetts: Addison-Wesley Publishing Company, 1977.
2. French and Vierck, GRAPHIC SCIENCE AND DESIGN, New York: McGraw-Hill Book Company, 1970.
3. Giesecke, Mitchell, Spencer, ENGINEERING GRAPHICS, New York: Collier-Macmillan.
4. Luzadder, Warren J., BASIC GRAPHICS (Second Edition), Englewood Cliffs, New Jersey: Prentice-Hall, Inc.
5. Beakley & Chilton, INTRODUCTION TO ENGINEERING DESIGN & GRAPHICS, New York: Collier-Macmillan, 1973.

Grade sheet

Probs.	Grades	Probs.	Grades	Probs.	Grades

WEEKLY TEST GRADES

TEST 1 ________

TEST 2 ________

TEST 3 ________

TEST 4 ________

TEST 5 ________

TEST 6 ________

TEST 7 ________

TEST 8 ________

TEST 9 ________

TEST 10 ________

TEST 11 ________

TEST 12 ________

TEST 13 ________

TEST 14 ________

TEST 15 ________

TEST 16 ________

TEST AVERAGE ________

MID-SEMESTER GRADE

DAILY AVERAGE ________

TEST AVERAGE ________

AVERAGE ________

FINAL GRADE

DAILY AVERAGE ________

TEST AVERAGE ________

FINAL GRADE ________

Grades

This graph can be used to compute grades for (1) daily assignments, (2) test grades, (3) written reports, (4) oral reports, (5) uniqueness, and (6) working drawing grades. Examples of each are shown below.

1. DAILY ASSIGNMENTS & TEST GRADES
   Assignments not turned in count as zeros; however, by doing extra assignments, your grades can be improved. The example below illustrates how a grade is improved when six extra assignments have been completed.

Number Assigned	30
Number Extra	6
TOTAL	36

AVERAGE GRADE FOR ALL 36 — 86

$$F = \frac{\text{No. completed} \times 100}{\text{No. assigned}} = \frac{36 \times 100}{30} = 120$$

From chart (F = 120) GRADE = 90

2. WRITTEN REPORTS, ORAL REPORTS, UNIQUENESS, WORKING DRAWINGS.
   Students should complete rating sheets from book for each of these areas and compute the F-values for each team member. These are used to compute each student's grade from the chart.

If the overall team grade is 86
the individual grades are:

Names (N=5)	C = %	F = CN	Graph
J. D. Doe	20%	100	86
H. P. Brown	16%	80	80
L. O. Smith	24%	120	90
R. L. Black	20%	100	86
T. O. Jones	20%	100	86

3. DAILY GRADES
   - Daily problems assigned ________
   - Extra daily problems ________
   - Total daily problems ________
   - Avg. grade on all probs. ________
   - GRADE FROM CHART ________

TEST GRADES
   - Number given ________
   - No. of extra assignments ________
   - Total test grades ________
   - GRADE FROM CHART ________

TEAM PROJECT (from chart)
   - Written report ________
   - Oral report ________
   - Uniqueness ________
   - Working drawings ________

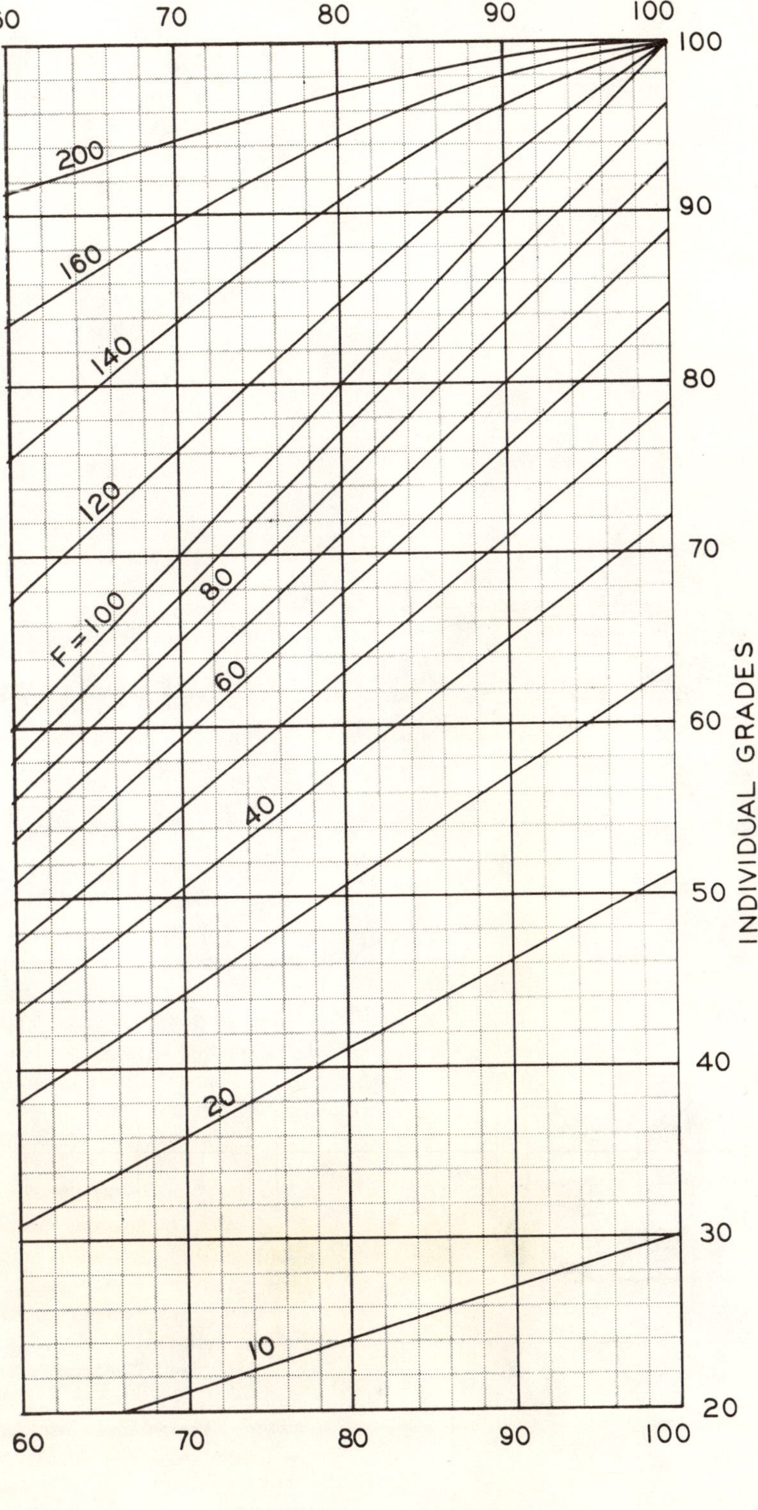

OVERALL AVERAGE [ ]

EXPECTED LETTER GRADE [ ]

Projects

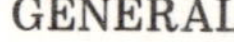

GENERAL

The problems given here are examples of team projects that can be assigned as course projects. The proper solution of these problems requires the application of graphical principles to develop, analyze, and present the final solution. You may be assigned these problems by your instructor with one or all of the requirements below:

Product Design:

1. Develop working drawings of your completed design.
2. Prepare a technical report that describes your product, where it will be sold, and the expected profits.
3. Build a working model of your completed design.

Systems Problem:

1. Prepare a technical report that describes the solution to your problem. Include the necessary graphs, diagrams, and layouts required to present your solution.

Both Types of Problems:

Give an oral class presentation within the assigned time limitation that will describe your solution and support your recommendations.

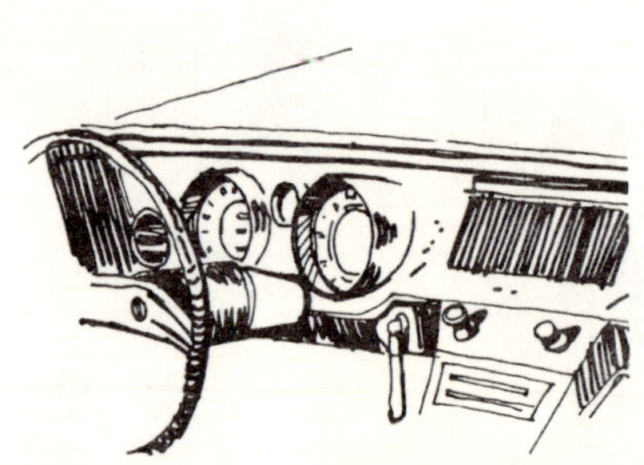

Product design

AUTO COFFEE

A manufacturer wishes to develop an attachment that can be installed in an automobile that will heat water from the existing manifold system. This water should be available at a tap near the dashboard for making instant coffee, tea or soup.

This apparatus should be easy to install, to maintain and to refill with fresh water. It should be economical in order to attract the consumer.

Adaptations of this design for other applications should be considered. Perhaps there are other more marketable uses for a similar device other than for making coffee or soup.

Prepare drawings and a model of your design as outlined in the design specifications. Make an estimate of the manufacturing costs, shipping weight, potential customers and other pertinent information.

PROJECTOR CABINET

Most classrooms are not designed to accommodate visual-aid equipment such as slide projectors and movie projectors. This equipment is carried into the classroom by the teacher and is plugged into an electrical outlet using extension cords. The projectors are usually placed on student desks, chairs or other make-shift positions that are not always the best for effective use.

It would be desirable to have a portable cabinet that would contain both a slide projector and a movie projector as a permanent fixture in a classroom. This cabinet should permit easy access to the projector by the teacher with the maximum of security. The enclosure should contain storage space for films and slide trays, an electrical system, and adjustments to allow the optimum effectiveness of projection.

A well-designed cabinet should have a broad market since every classroom could be considered as a possible market for this product.

Prepare drawings and model of your design as outlined in the design specifications. Make an estimate of the manufacturing costs, shipping weight, potential customers and other pertinent information.

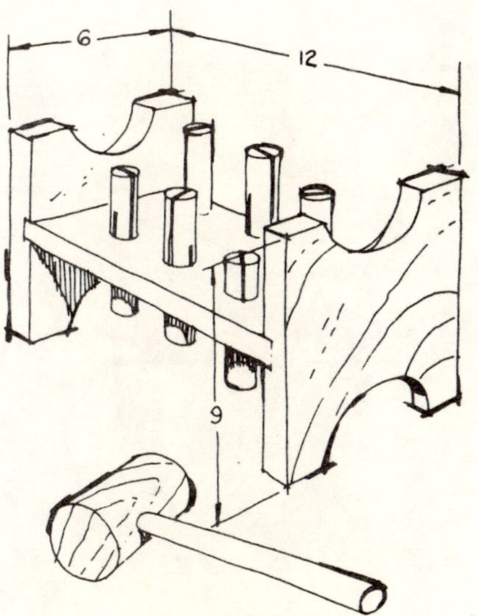

Systems design

MANUFACTURING OPERATION FOR A TOY

An example of an eductional toy is shown in the illustration. This toy is made entirely of wood and is held together with screws. Your retailers have indicated that they expect to sell 5,000 of these per month with expectations of increasing sales after six months. With these prospects it appears to be feasible to consider establishing a production system for manufacturing these toys.

Using either this design or a design selected or designed by your team, develop plans for the manufacturing facility necessary for producing these toys. Determine the square footage needed, types of machines required, space for raw materials, office facilities, and warehousing space needed to operate the installation. Determine the number of people that would be necessary, their rate of pay, and the number of work stations required. Your completed project should cover all aspects of the manufacturing process and the solution of the management problems required for an efficient and economical operation to produce the product.

Compute the expense to produce each item and the selling price necessary to provide the required profit margin.

CHILDREN'S ENTERTAINMENT CENTER

Today's housewife must make frequent visits to shopping centers to purchase household items and groceries. Usually it is necessary that she take her young children with her during these shopping trips. This introduces a problem inasmuch as she must watch the child closely and even discipline the child from time to time which may distract her from her chores. It would be beneficial if shopping centers could either be designed or modified to provide children's entertainment centers where the children could play, take rides, or watch movies at a small cost to free their mothers during shopping trips.

Determine the specific needs for a children's entertainment center; the age of the children to be involved; the numbers of children; types of entertainment enjoyed by children; and similar factors of this type that would affect the solution. Perhaps the entertainment center could be established in the shopping center's parking lot at a minimum of expense. Another solution might be a separate building within the shopping center complex.

Scheduling

design schedule & progress record

TEAM: 2 PROJECT: PRELIMINARY DESIGN

WORK PERIODS: 4 MAN HRS: 20 FINISH DATE: 10-2

JOB	ASSIGNMENT	ESTIMD HOURS	ACTUAL HOURS	PER CENT COMPLETE 0 50 100
1	TABULATE QUESTION.	1.0	1.25	
2	WRITE VENDORS	2.0	1.30	
3	PREPARE APPENDIX	1.0	1.00	
4	BEGIN PROB IDEN.	0	0	
5	PREPARE QUESTION.	1.5		
6	GRAPH DATA	3.0		
7	VISIT LIBRARY	2.0		
8	ADMINISTER QUEST.	2.0		
9	DRAFT PROB IDEN.	2.0		

The technique below is suggested as a method of scheduling activities for your team project. Good scheduling is highly essential when working as a a team.

1. List jobs that must be done as they come to mind on the DESIGN SCHEDULE & PROGRESS RECORD. These need not be listed in order, but as you think of them. Estimate the number of man-hours required to complete each job and list in the column provided. The total should not exceed the total number of hours available.

2. These jobs should then be arranged in sequence by using an ACTIVITIES NETWORK. Notice that the numbers assigned to the jobs on the DESIGN SCHEDULE & PROGRESS RECORD are the same numbers used in the ACTIVITIES NETWORK. This gives you a graphical picture of the sequence in which the jobs should be performed.

3. List the jobs on the ACTIVITIES SEQUENCE CHART, taking them from the ACTIVITIES NETWORK, to arrive at a bar graph showing when each job should be performed. Notice that the same numbers are used on the ACTIVITIES SEQUENCE CHART as were used in the previous two steps. Your ACTIVITIES SEQUENCE CHART can be used for assigning jobs to your team members. Per cent completion should be graphed on the DESIGN SCHEDULE & PROGRESS RECORD as the jobs progress toward completion.

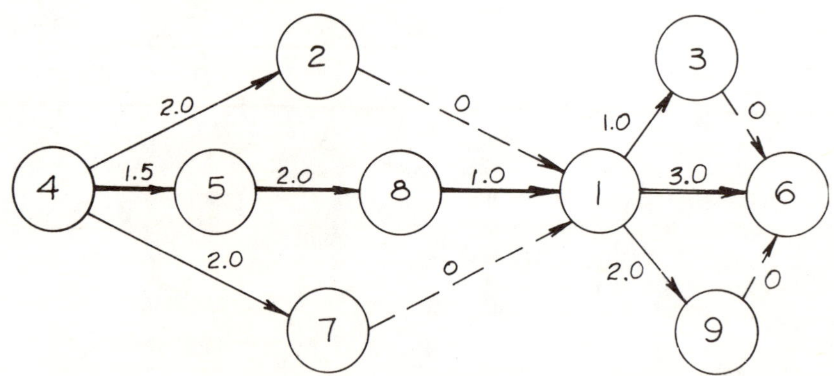

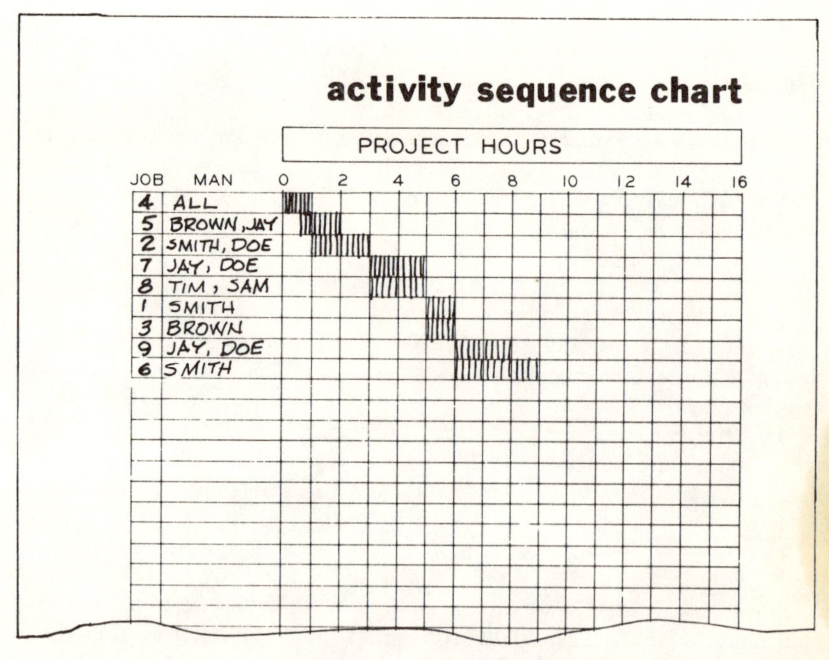

Design Schedule & Progress Record

TEAM ______ PROJECT ______________________________

WORK PERIODS ______ MAN-HOURS ______ FINISHING DATE _____ SECTION ______

JOB	ASSIGNMENT	ESTIMD HOURS	ACTUAL HOURS	PER CENT COMPLETE 0 50 100

Activity Sequence Chart

PROJECT HOURS

JOB	MAN	0	2	4	6	8	10	12	14	16

Progress Report

REPORT NUMBER ______

TEAM ______ PROJECT ______________________________

SECTION ______ SUBMITTED BY ______________________ DATE ______________

	0%	25%	50%	75%	100%
SCHEDULING					
List jobs					
Select team leader					
Prepare activity sequence chart					
Make individual assignments					
Establish work guidelines					
PROBLEM IDENTIFICATION					
Write problem statement					
Determine survey data needed					
Write letters to vendors					
Estimate market needs					
Identify physical characteristics: sizes weights, areas, etc.					
Gather library data					
Make job assignments					
Write draft for written report					
PRELIMINARY IDEAS					
Brainstorming session					
Make idea sketches					
Adapt similar design features					
Write draft for written report					
REFINEMENT					
Prepare scale drawings					
Determine angles, lengths, weights, sizes, and etc.					
Evaluate refined designs					
Write draft for written report					
ANALYSIS					
Evaluate merits of designs					
Cost analysis					
Market analysis					
Weight and size analysis					
Strength analysis					
Human engineering analysis					
Overhead expense analysis					
Write draft for written report					
DECISION					
Select best design to implement					
Compare aspects of each design					
Make final decision					
IMPLEMENTATION					
Complete working drawing					
Graph data for report					
Prepare visuals for oral report					
Type final report					
Prepare final oral report					
Complete working model					

Written Report

The table below should be completed jointly by the team with only the grade column being completed by the instructor who will use the chart on page 4 and the factor F that was computed for each team member.

TEAM NO. ______ PROJECT ______________________________

	NAMES	NO. (N =)	% CONTRIBUTION (C)	F = NC	GRADE (G)
1.					
2.					
3.					
4.					
5.					
6.					
7.					
8.					

	EVALUATION BY INSTRUCTOR 100%	Max. Value	Points Earned
1.	Use of an appropriate cover	2	____
2.	Inclusion of an evaluation sheet	2	____
3.	Inculsion of a proper letter of transmittal	2	____
4.	Correct title page	2	____
5.	Proper table of contents	2	____
6.	Sufficient introduction to the the report	5	____
7.	Thoroughness in identifying the problem	10	____
8.	Continuity and quality of the body the report	10	____
9.	Collection and presentation of background data	5	____
10.	Justification of major decisions	5	____
11.	Review of costs, overhead expenses, shipping costs and similar expenses	5	____
12.	Arrival at strong conclusion and recommendation	5	____
13.	Sufficient number of graphs and graphics	10	____
14.	Quality of graphics	10	____
15.	Bibliography—form and content	5	____
16.	Use of footnotes	5	____
17.	Appendix—content and form	5	____
18.	Form and appearance of report (spelling, punctuation, margins, typing, neatness)	10	____
		100	

Oral Report

The table below should be completed jointly by the team with only the grade column being completed by the instructor who will use the chart on page 4 and the factor F that was computed for each team member.

TEAM NO. ______ PROJECT ______________________________

	NAMES	NO. (N =)	% CONTRIBUTION (C)	F = NC	GRADE (G)
1.	______		______	______	______
2.	______		______	______	______
3.	______		______	______	______
4.	______		______	______	______
5.	______		______	______	______
6.	______		______	______	______
7.	______		______	______	______
8.	______		______	______	______

EVALUATION BY INSTRUCTOR 100%	Max. value	Points earned
1. Introduction of team members	2	______
2. Statement of purpose of the presentation	5	______
3. Continuity of presentation	3	______
4. Use of visuals — point to important points, do not block screen, do not fumble, etc.	10	______
5. Clear presentation of recommended design	10	______
6. Presentation of alternative solutions considered	2	______
7. Coverage of economics — manufacturing, shipping packing, overhead, mark-up, etc.	10	______
8. Consideration of human factors	5	______
9. Presentation of an effective conclusion	5	______
10. Poise and professionalism	2	______
11. Proper dress for the presentation	2	______
12. Use of adequate number of visual aids	9	______
13. Quality of visual aids includeing model	15	______
14. Participation of team members (perfect score if all participate)	10	______
15. Use of allotted time	10	______
	100	

Additional comments by the instructor on the back of this sheet.

The table below should be completed jointly by the team with only the grade column being completed by the instructor who will use the chart on page 4 and the factor F that was computed for each team member.

Project Uniqueness

TEAM NO. _____ PROJECT ______________________________

	NAMES NO. (N =)	% CONTRIBUTION (C)	F = NC	GRADE (G)
1.				
2.				
3.				
4.				
5.				
6.				
7.				
8.				
		100%		

EVALUATION BY INSTRUCTOR: Degree of uniqueness and originality in solution.

		Max. Value	Points Earned
1.	Serves a needed function	10	____
2.	Functions effectively	10	____
3.	Needed by consumers	10	____
4.	Simple, uncomplicated solution	10	____
5.	Reasonable, attractive price	10	____
6.	An attractive investment for marketing	10	____
7.	Level of imagination and ingenuity	10	____
8.	Aesthetically pleasing and attractive	10	____
9.	Competition of similar products on the market	10	____
10.	Degree of fulfillment of problem statement	10	____
		100	

Additional comments by the instructor are on the back of this sheet.

Working Drawings

TEAM NO. _____ PROJECT ______________________________

NAMES	NO. (N =)	% CONTRIBUTION (C)	F = NC	GRADE (G)
1.				
2.				
3.				
4.				
5.				
6.				
7.				
8.				

EVALUATION BY INSTRUCTOR 100%		Max. Value	Points Earned
TITLE BLOCK (5 Points)			
Student's name	1		
Checker	1		
Date	1		
Scale	1		
Sheet number	1		
ORTHOGRAPHIC DETAILS (19 points)			
Proper views	10		
Spacing of views	5		
Part names and numbers	2		
Correct sections & conventions	2		
DESIGN INFORMATION (12 Points)			
Tolerances	5		
Finish marks	2		
Thread notes	5		
DIMENSIONING (20 Points)			
Proper arrowheads	3		
Spacing of dimension lines	3		
Adequate dimensions	4		
Fillets and rounds noted	2		
Hole notes	2		
Inch marks omitted	2		
Dimensions from best views	4		
DRAFTSMANSHIP (19 Points)			
Line weights	6		
Lettering	6		
Neatness	3		
Reproduction quality	4		
GENERAL NOTES (5 Points)			
SI symbol	2		
Third angle symbol	2		
General tolerance note	1		
ASSEMBLY (15 Points)			
Descriptive views	6		
Clarity	3		
Parts list	4		
Part numbers	2		
PRESENTATION (5 Points)			
Stapled	2		
Trimmed	1		
Folded	1		
Grade sheet attached	1		
Total	100		

LETTERING
VERTICAL CAPITALS

USING AN F OR HB PENCIL WITH A SLIGHTLY ROUNDED POINT, CONSTRUCT EACH LETTER IN THE SPACES PROVIDED. OBSERVE THE FORM AND PROPORTION OF EACH LETTER TO ASSIST YOU IN IMPROVING YOUR LETTERING IN FUTURE ASSIGNMENTS.

NAME
FILE SEC DATE
MIN. GRADE

METRIC SCALES

Remove this page and fold it longwise along the desired scale for making metric measurements.

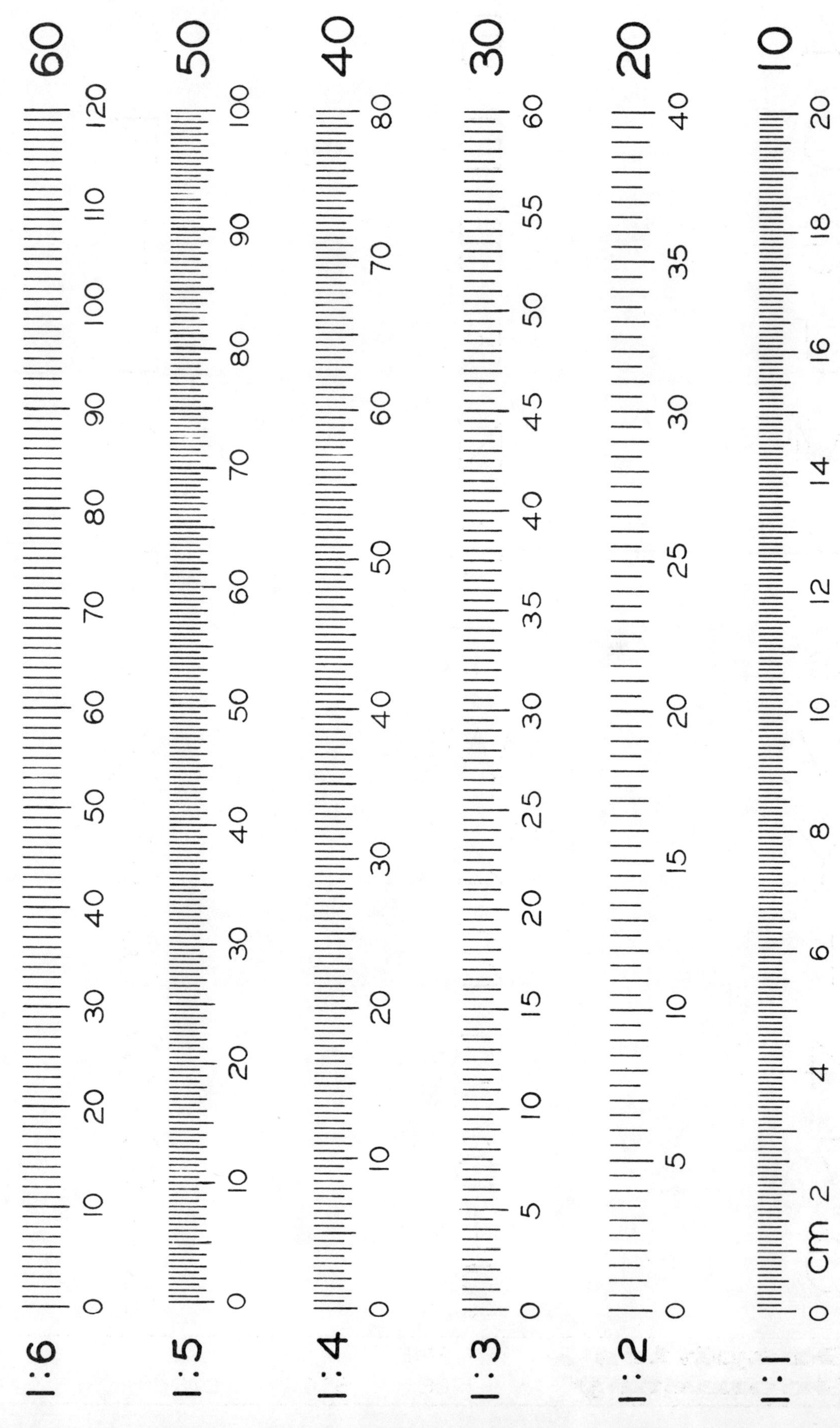

LETTERING

USING AN F OR HB PENCIL WITH A SLIGHTLY ROUNDED POINT, CONSTRUCT EACH LETTER IN THE SPACES PROVIDED. OBSERVE THE FORM AND PROPORTION OF EACH LETTER TO ASSIST YOU IN IMPROVING FUTURE ASSIGNMENTS. THESE ARE LOWER CASE LETTERS.

USING LOWER CASE VERTICAL LETTERING, REPEAT THE FOLLOWING SENTENCE: "ENGINEERING DRAWING IS THE LANGUAGE OF INDUSTRY THROUGH WHICH ALL DESIGNS ARE DEVELOPED." USE THE GUIDE LINES PROVIDED.

METRIC SCALES

Remove this page and fold it longwise along the desired scale for making metric measurements.

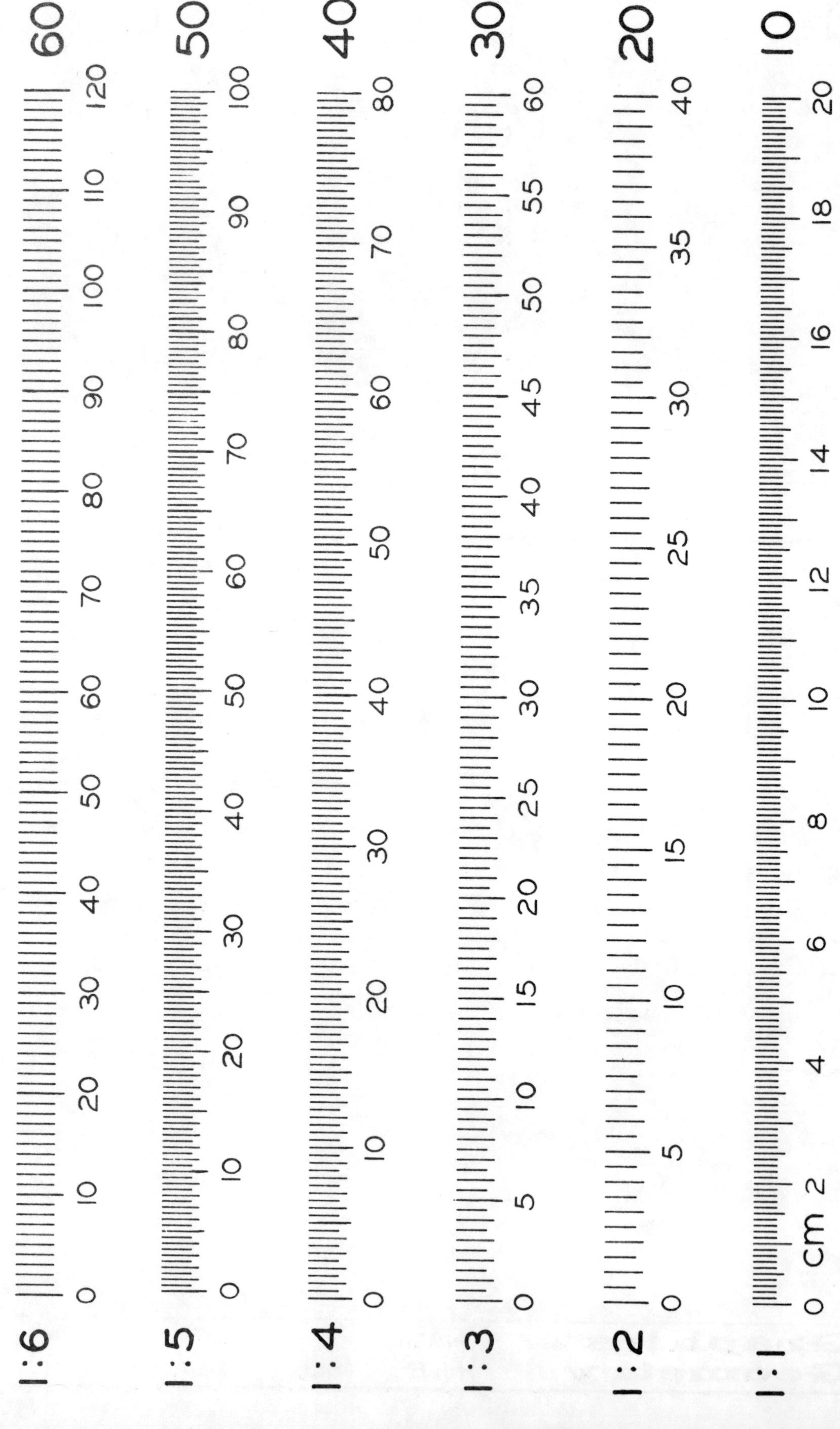

LETTERING

USING AN F OR HB PENCIL WITH A SLIGHTLY ROUNDED POINT, CONSTRUCT EACH LETTER IN THE SPACES PROVIDED. OBSERVE THE FORM AND PROPORTION OF EACH LETTER TO ASSIST YOU IN IMPROVING YOUR LETTERING IN FUTURE ASSIGNMENTS.

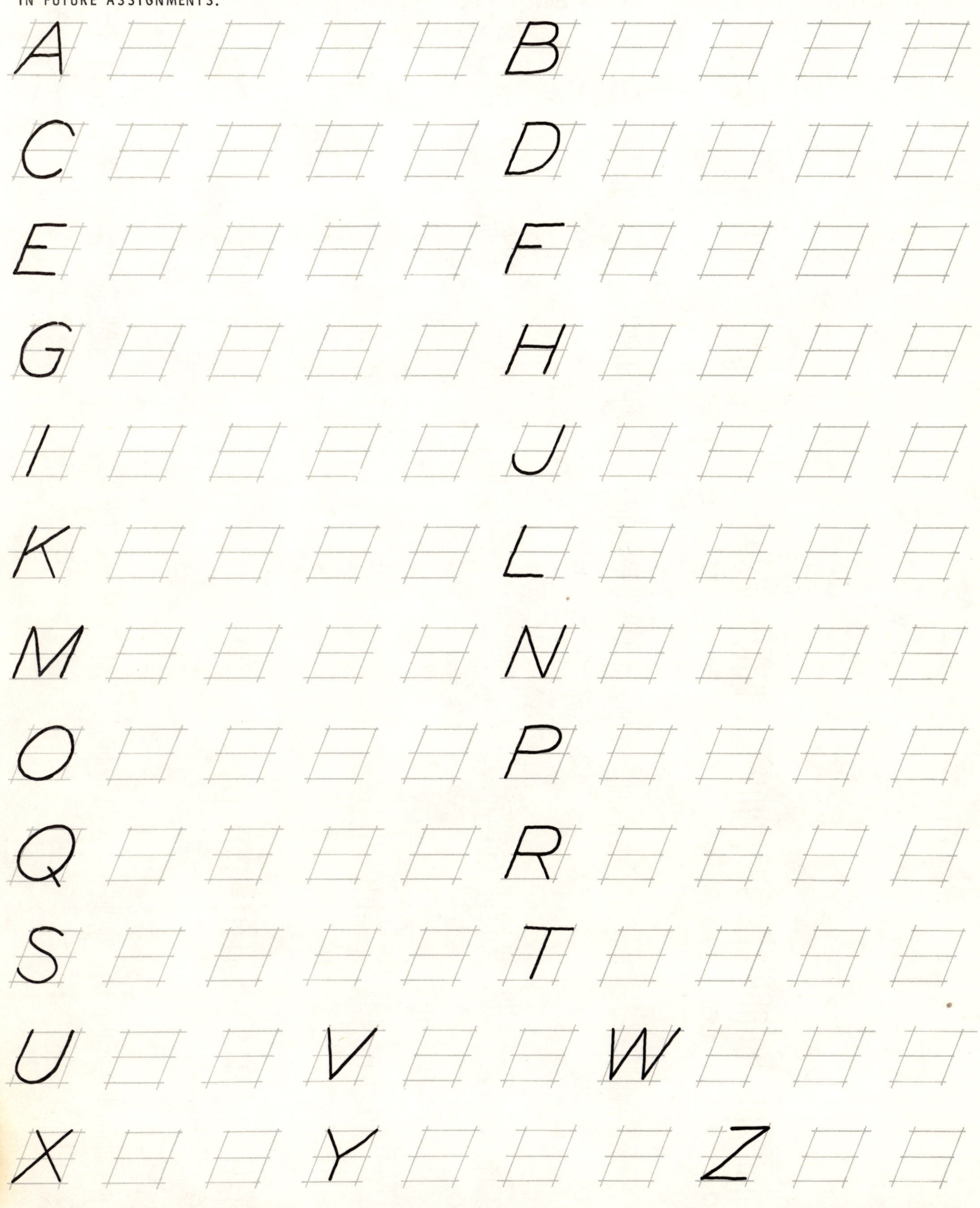

Graphics & Geometry ©	NAME FILE SEC DATE	MIN.	GRADE	17

METRIC SCALES

Remove this page and fold it longwise along the desired scale for making metric measurements.

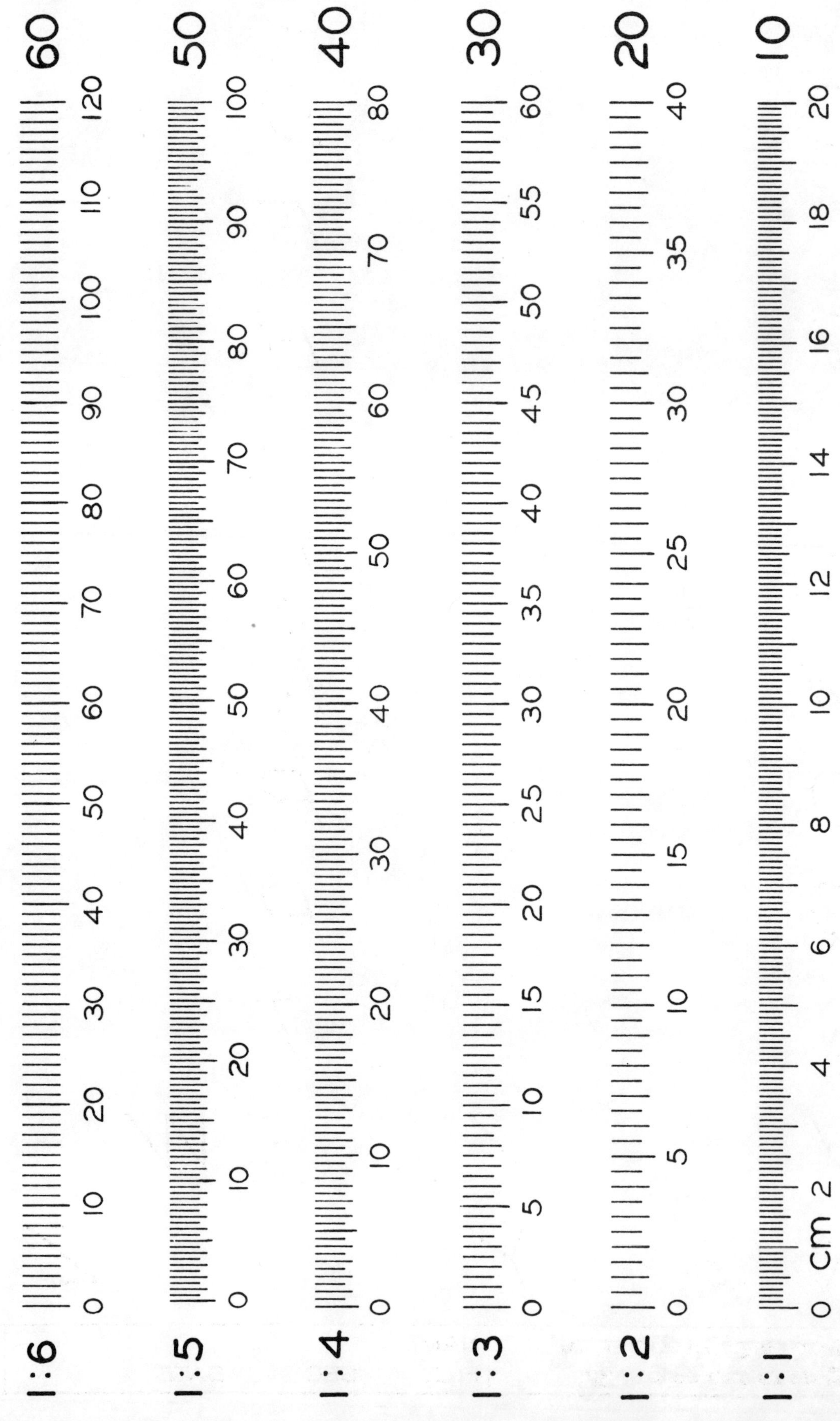

USE OF INSTRUMENTS

A COMMUNICATIONS TRANSMITTER LIKE THE ONE SHOWN AT THE RIGHT UTILIZES A "CARRIER" FREQUENCY GENERATED BY AN OSCILLATOR CIRCUIT SIMILAR TO THE ONE SKETCHED BELOW. MAKE A DOUBLE-SIZE INSTRUMENT DRAWING OF THE SKETCHED CIRCUIT AND LETTER THE COMPONENTS AS SHOWN. START IN THE UPPER LEFT CORNER AT POINT 'A' AND USE THE FULL-SIZE CIRCUIT COMPONENT SYMBOLS GIVEN. MAKE THICK SHARP BLACK LINES FOR ALL SYMBOLS AND LINE WORK.

COURTESY OF THE HEATH CO.

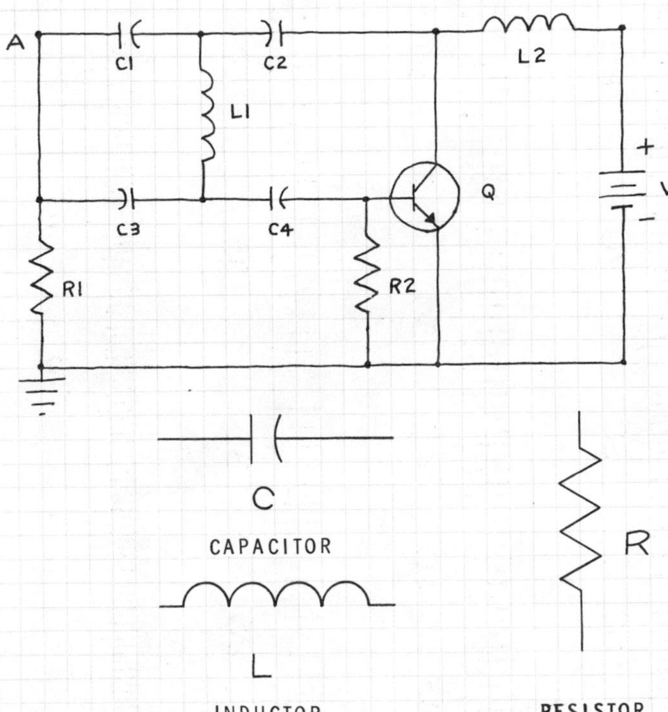

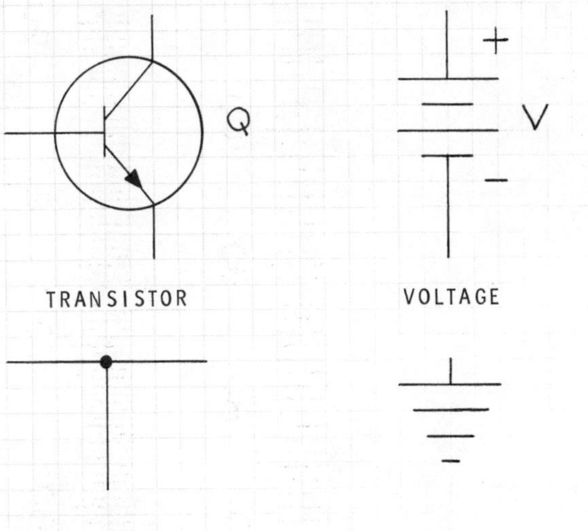

TRANSISTOR

VOLTAGE

C

CAPACITOR

L

INDUCTOR

R

RESISTOR

CONNECTION

GROUND

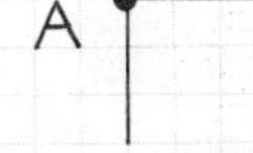

REB

METRIC SCALES

Remove this page and fold it longwise along the desired scale for making metric measurements.

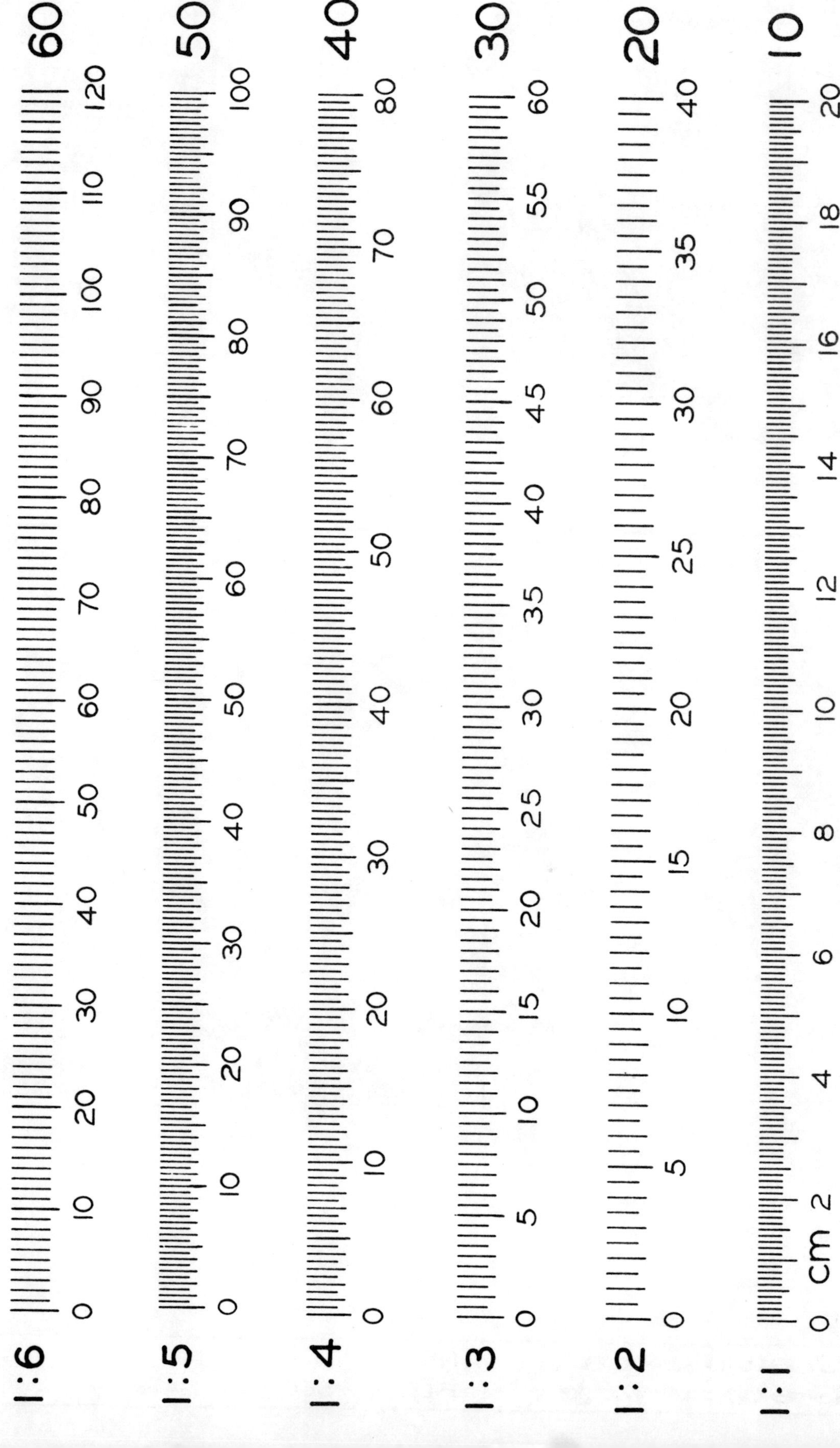

ARCHITECTS' SCALE

MEASURE LENGTHS A, B, AND C ON FIGURE 1 USING THE SCALES LISTED BELOW. LETTER EACH DIMENSION IN THE APPROPRIATE COLUMN OPPOSITE THE SCALE USED. MAKE EACH MEASUREMENT TO THE NEAREST SMALL DIVISION ON THE SCALE.

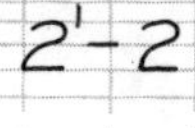
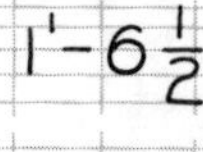

SCALE	A	B	C
EX. 1 = 1'-0	2'-2	1'-6 $\frac{1}{2}$	2'-7 $\frac{3}{4}$
1. $\frac{1}{2}$ = 1'-0			
2. $\frac{3}{4}$ = 1'-0			
3. $\frac{3}{8}$ = 1'-0			
4. 1 $\frac{1}{2}$ = 1'-0			
5. $\frac{1}{4}$ = 1'-0			

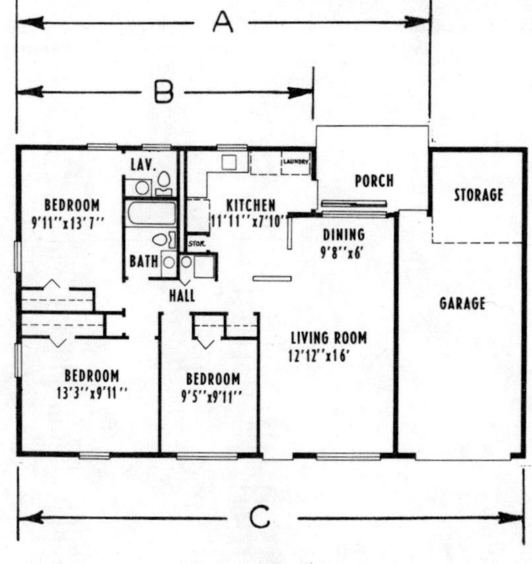

FIGURE 1

ENGINEERS' SCALE

MEASURE LENGTHS D, E, AND F ON FIGURE 2 USING THE SCALES LISTED BELOW. LETTER EACH DIMENSION IN THE APPROPRIATE COLUMN OPPOSITE THE SCALE USED. MAKE EACH MEASUREMENT TO THE NEAREST HALF OF THE SMALLEST SCALE DIVISION.

SCALE	D	E	F
EX. 1=10.0'	20.5'	48.0'	27.0'
6. 1=30.0'			
7. 1=400'			
8. 1=50.0'			
9. 1=20.0'			
10. 1=600'			

ANGLES

MEASURE THE INTERIOR AND DEFLECTION ANGLES AT CORNERS 1, 2, 3, AND 4 ON FIGURE 2. LETTER EACH ANGLE IN THE APPROPRIATE COLUMN BELOW. MAKE EACH MEASUREMENT TO THE NEAREST DEGREE IN A CLOCKWISE DIRECTION. TOTAL BOTH COLUMNS.

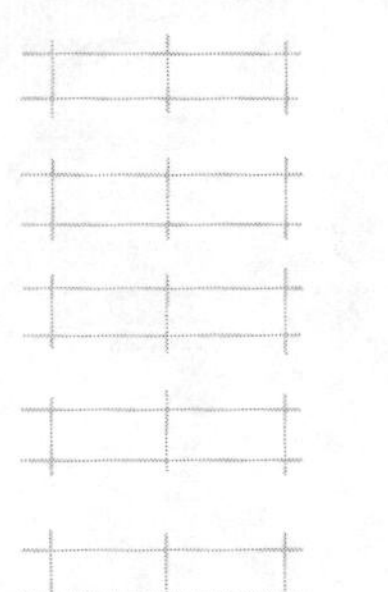
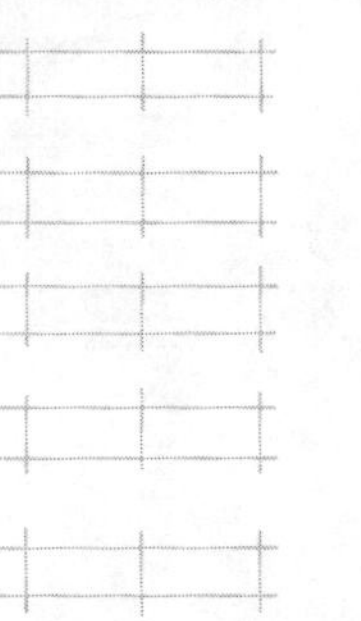

	INTERIOR ∠'S	DEFLECTION ∠'S
11. 1		
12. 2		
13. 3		
14. 4		
15. TOTAL		

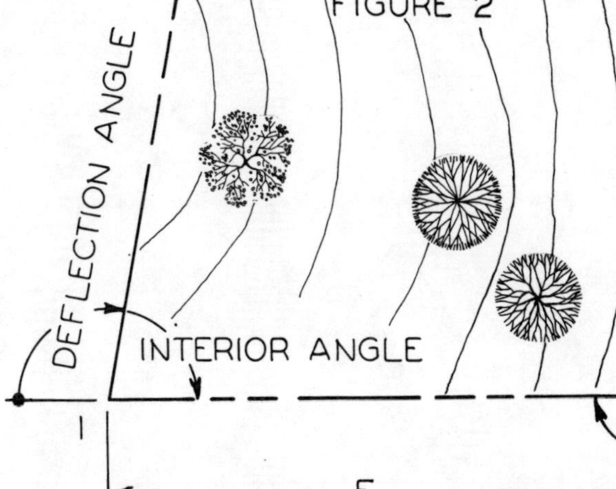

FIGURE 2

REB

METRIC SCALES

Remove this page and fold it longwise along the desired scale for making metric measurements.

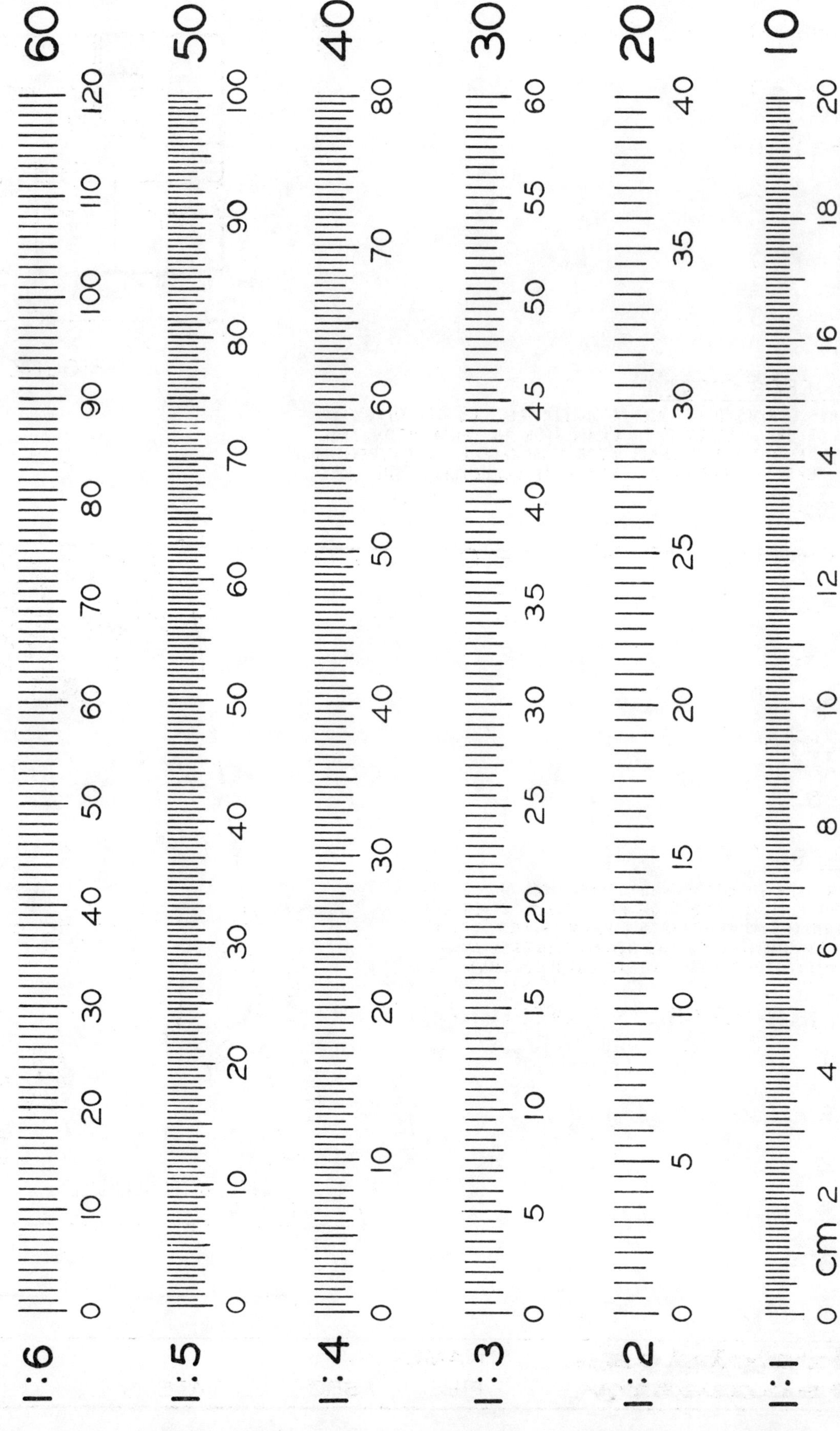

METRIC SCALES

MEASURE THE DIMENSIONS ON THE PART (FIG. 1), USING THE METRIC SCALES INDICATED. LETTER THE VALUES UNDER THE HEADINGS: A, B, & C. THE DIMENSIONS UNDER A WILL BE IN MILLIMETERS, UNDER B IN CENTIMETERS AND UNDER C IN METERS.

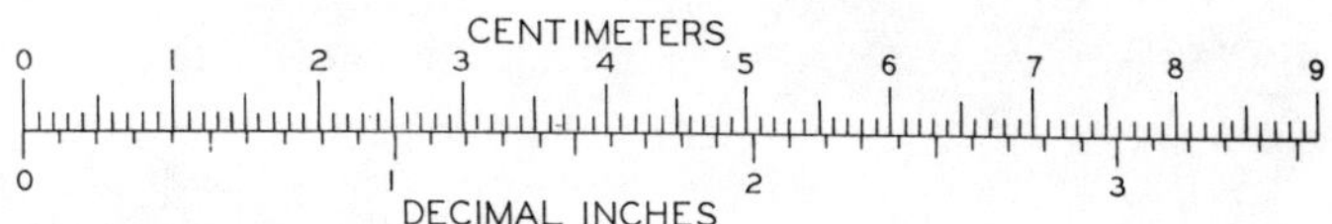

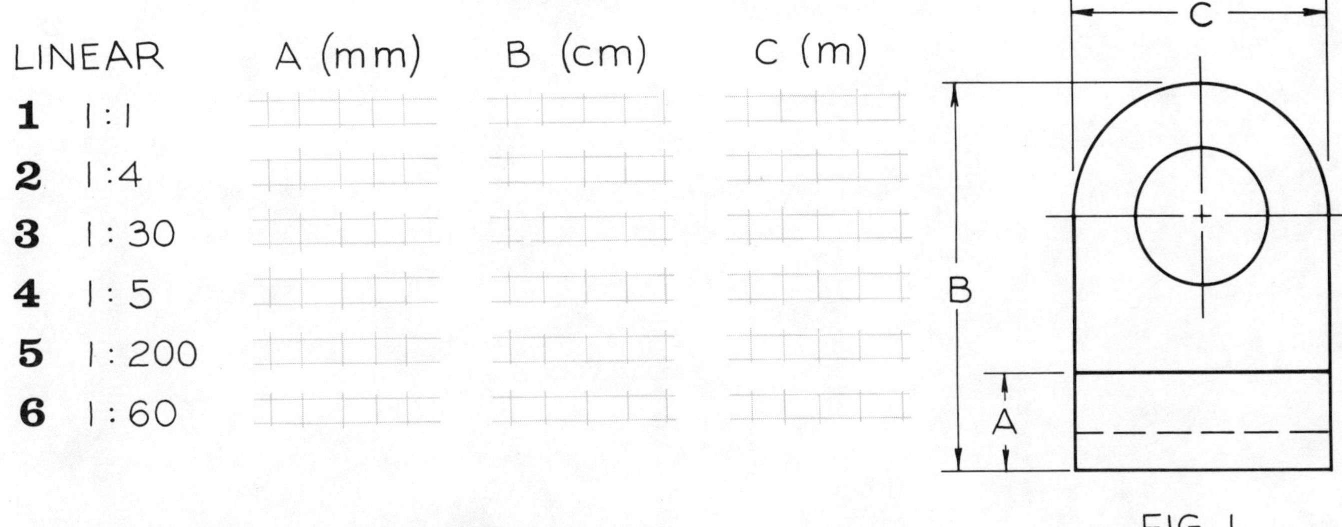

LINEAR		A (mm)	B (cm)	C (m)
1	1:1			
2	1:4			
3	1:30			
4	1:5			
5	1:200			
6	1:60			

FIG. 1

USING MILLIMETERS AS THE UNIT OF MEASUREMENT, DETERMINE THE DIMENSIONS OF THE PARALLELOGRAM AND FIND ITS AREA FOR EACH SCALE INDICATED.

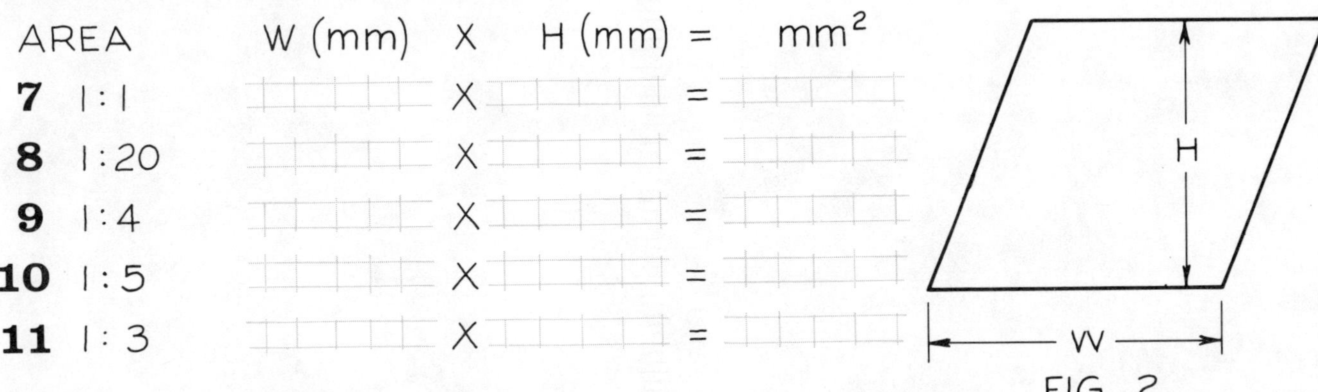

AREA		W (mm)	X	H (mm)	=	mm^2
7	1:1		X		=	
8	1:20		X		=	
9	1:4		X		=	
10	1:5		X		=	
11	1:3		X		=	

FIG. 2

FIND THE VOLUME OF THE PARALLELOGRAM IN FIG. 2 IF ITS THICKNESS VARIES AS GIVEN. EXPRESS THE VOLUMES IN LITERS. 1 LITER = 1000 CUBIC CENTIMETERS.

MASS		AREA (W·H)	X THICKNESS	X VOLUME	= LITERS
		(mm^2)	X (mm)	= (mm^3)	=
11	1:1		X 10	=	=
12	1:20		X 17	=	=
13	1:4		X 30	=	=
14	1:5		X 40	=	=
15	1:3		X 60	=	=

JE

JHE

Graphics & Geometry ©	NAME FILE SEC DATE	MIN.	GRADE	20

METRIC SCALES

1

THE MILLIMETER IS THE STANDARD UNIT OF MEASUREMENT THAT IS USED TO DIMENSION AN ENGINEERING DRAWING. MEASURE AND SUPPLY THE MISSING DIMENSIONS ON THE WORKING DRAWING BELOW. THIS DRAWING IS DRAWN HALF-SIZE. USE THE GUIDELINES THAT ARE GIVEN.

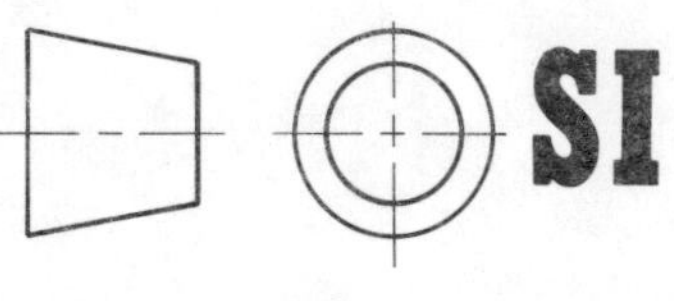

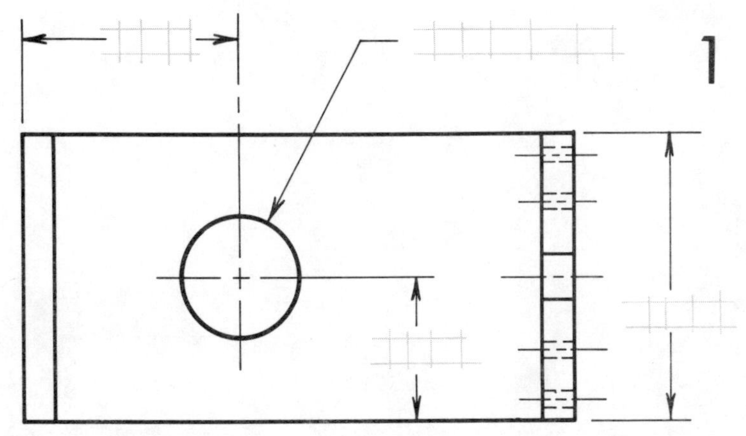

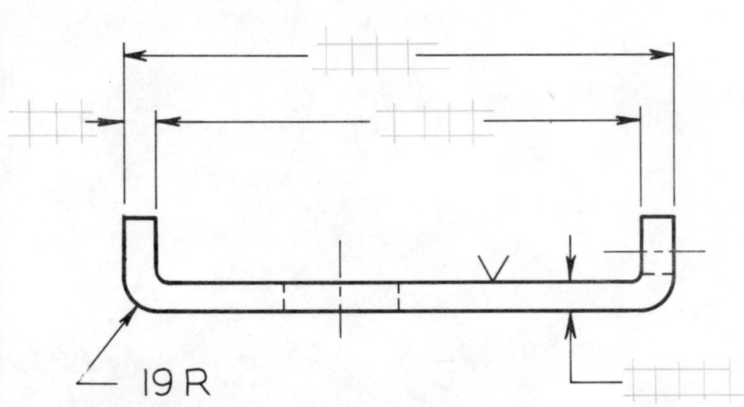

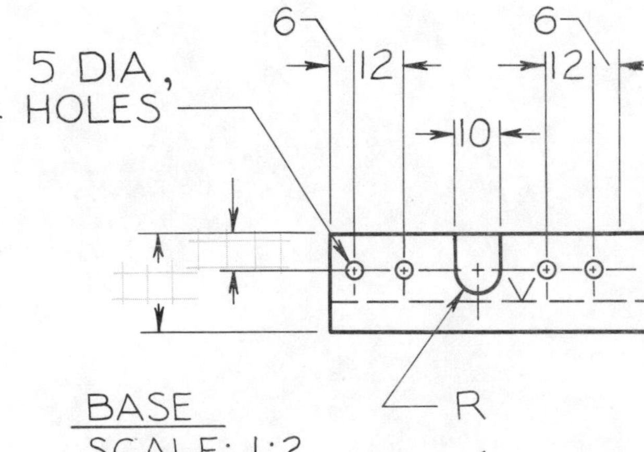

BASE
SCALE: 1:2

THREE SCALES ARE GIVEN BELOW WHICH ARE CALIBRATED WITH ENGLISH UNITS OF MEASUREMENT. THE CORRESPONDING END VALUES ARE GIVEN IN METRIC UNITS. COMPLETE THE CALIBRATIONS ON THE UPPER SIDE OF EACH SCALE TO FORM AN ALIGNMENT SCALE THAT CONVERTS ONE SET OF VALUES INTO THE OTHER. USE THE DIAGONAL LINE TECHNIQUE ILLUSTRATED AT THE RIGHT. LABEL THE CALIBRATIONS.

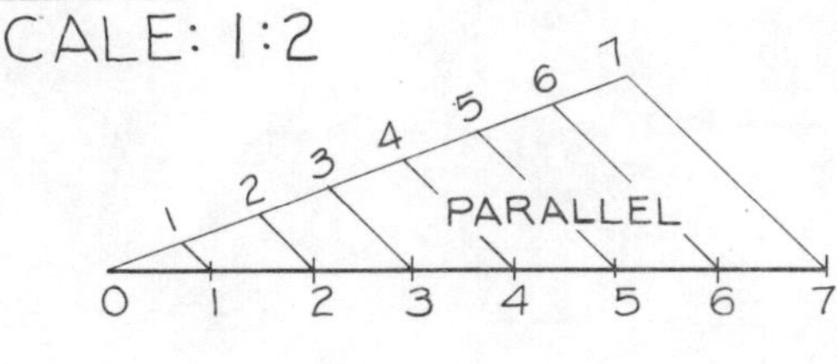

2

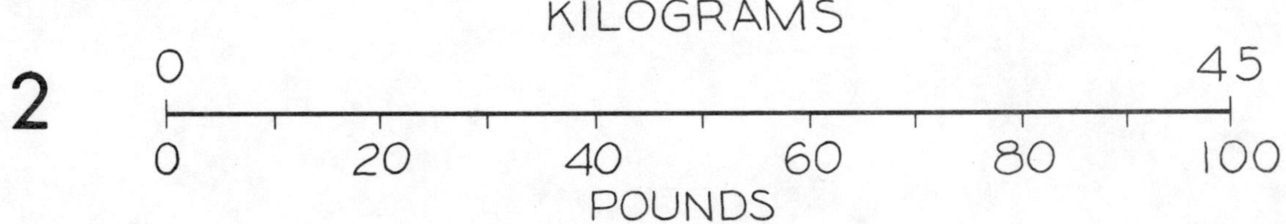

3

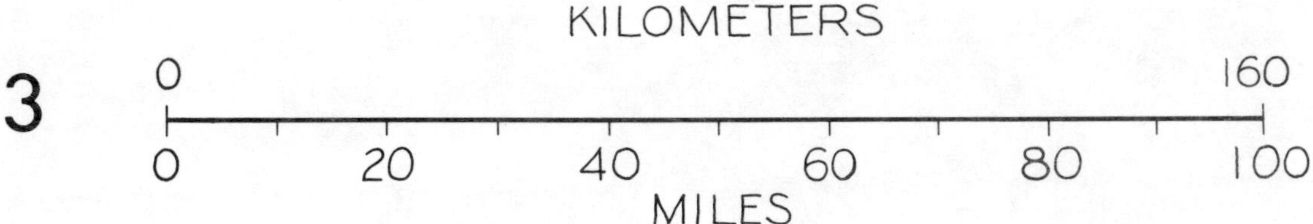

4

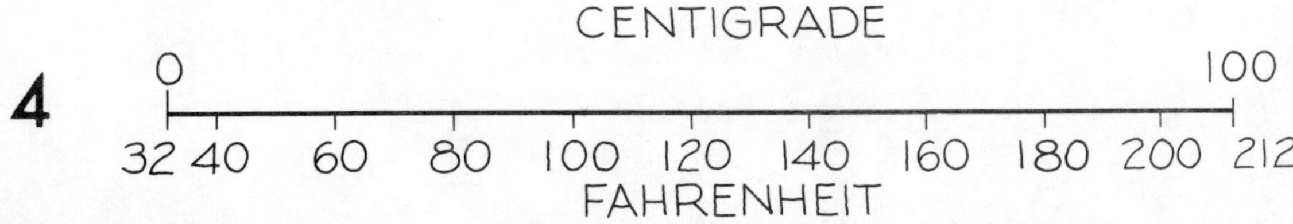

JHE

Graphics & Geometry ©	NAME FILE SEC DATE	MIN.	GRADE	21

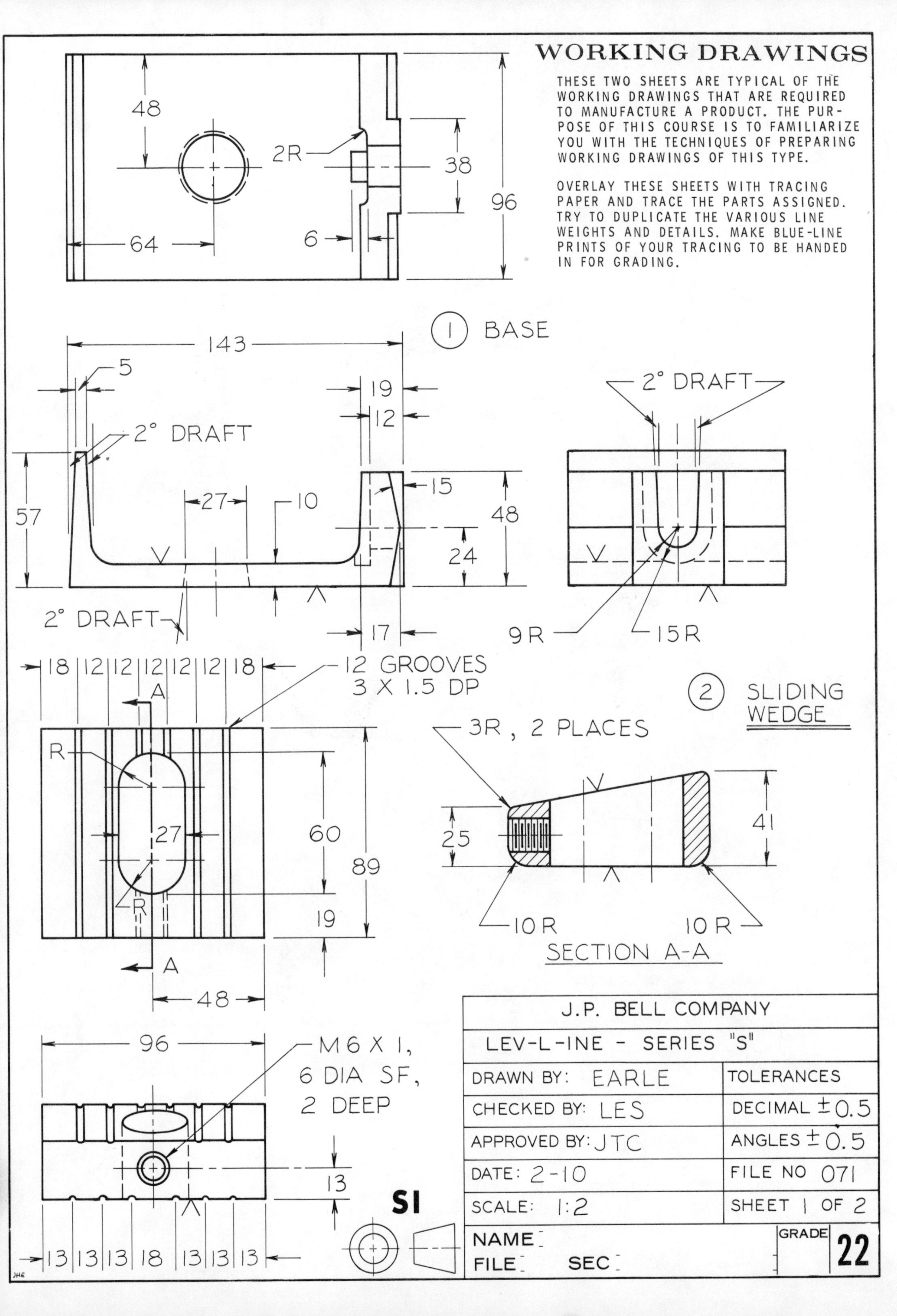
WORKING DRAWINGS
THESE TWO SHEETS ARE TYPICAL OF THE WORKING DRAWINGS THAT ARE REQUIRED TO MANUFACTURE A PRODUCT. THE PURPOSE OF THIS COURSE IS TO FAMILIARIZE YOU WITH THE TECHNIQUES OF PREPARING WORKING DRAWINGS OF THIS TYPE.
OVERLAY THESE SHEETS WITH TRACING PAPER AND TRACE THE PARTS ASSIGNED. TRY TO DUPLICATE THE VARIOUS LINE WEIGHTS AND DETAILS. MAKE BLUE-LINE PRINTS OF YOUR TRACING TO BE HANDED IN FOR GRADING.
48
2R
38
96
64
6
1 BASE
143
5
19
12
2° DRAFT
57
27
10
15
48
24
2° DRAFT
17
2° DRAFT
9 R
15 R
18 12 12 12 12 12 18
12 GROOVES
3 X 1.5 DP
A
2 SLIDING WEDGE
R
3R, 2 PLACES
27
60
89
25
41
R
19
10 R
10 R
SECTION A-A
A
48
J.P. BELL COMPANY
LEV-L-INE - SERIES "S"
96
M 6 X 1,
6 DIA SF,
2 DEEP
DRAWN BY: EARLE
TOLERANCES
CHECKED BY: LES
DECIMAL ± 0.5
APPROVED BY: JTC
ANGLES ± 0.5
DATE: 2-10
FILE NO 071
13
SI
SCALE: 1:2
SHEET 1 OF 2
NAME
GRADE
22
FILE
SEC
13 13 13 18 13 13 13

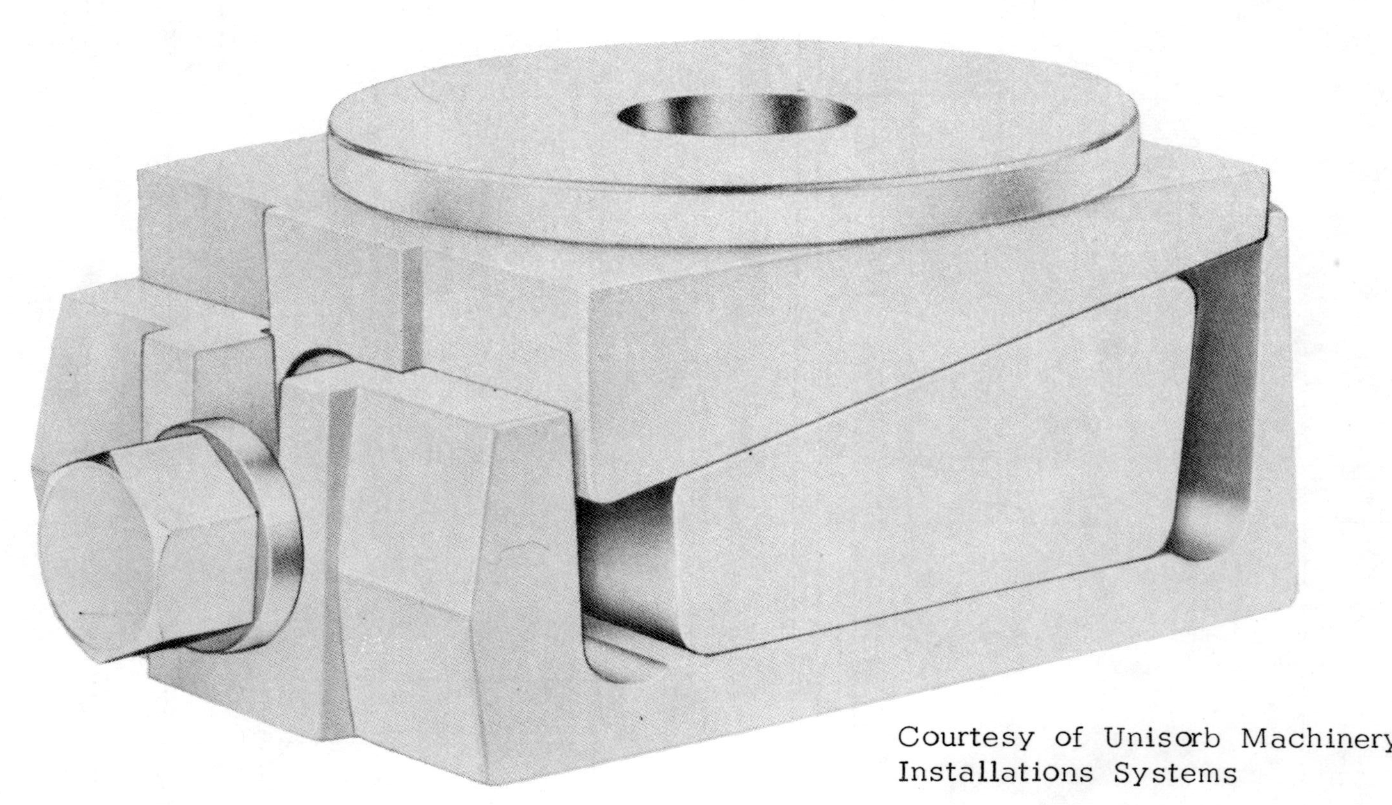

Courtesy of Unisorb Machinery
Installations Systems

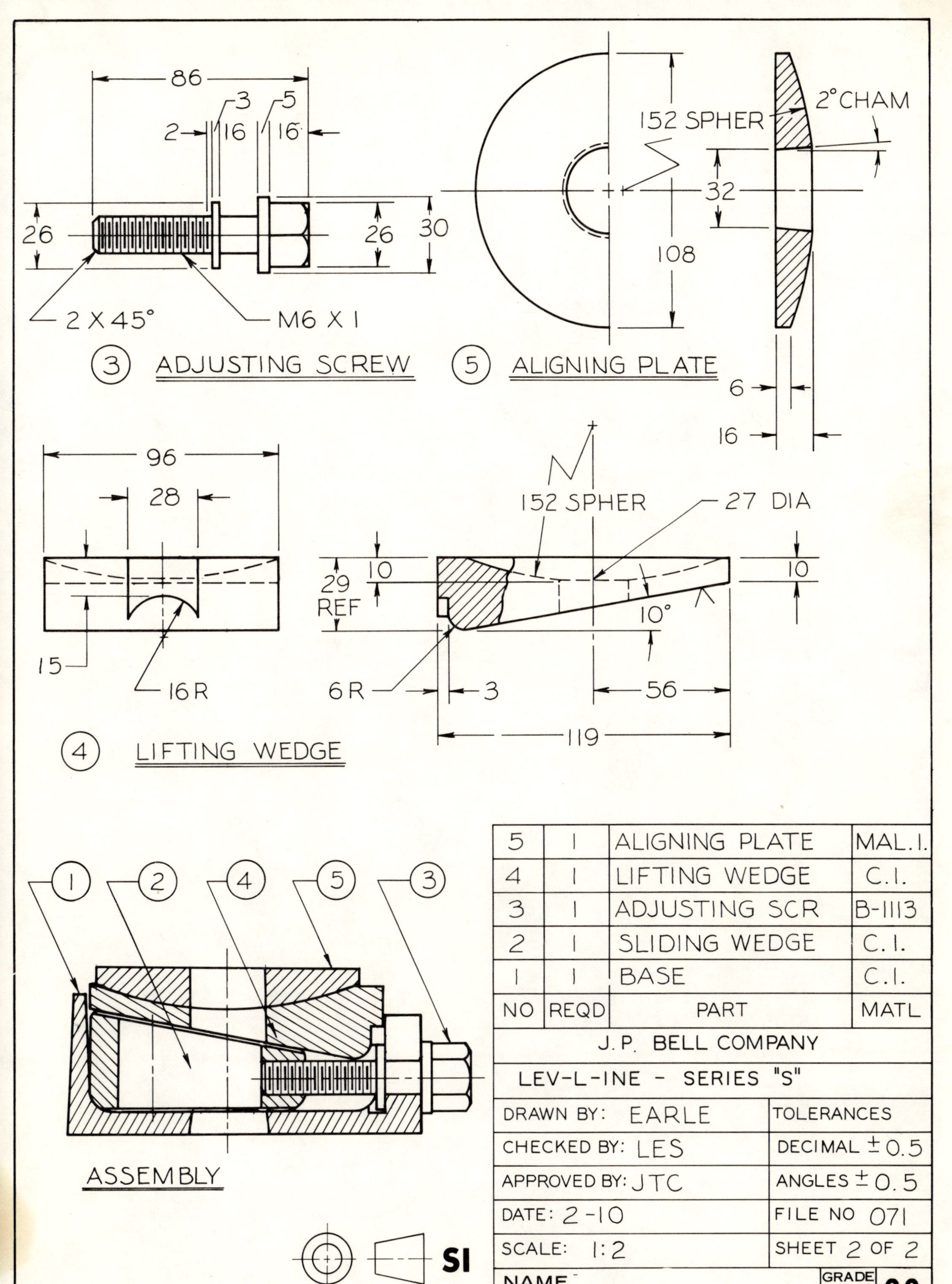

5	1	ALIGNING PLATE	MAL. I.
4	1	LIFTING WEDGE	C. I.
3	1	ADJUSTING SCR	B-1113
2	1	SLIDING WEDGE	C. I.
1	1	BASE	C. I.
NO	REQD	PART	MATL

J. P. BELL COMPANY

LEV-L-INE - SERIES "S"

DRAWN BY: EARLE	TOLERANCES
CHECKED BY: LES	DECIMAL ± 0.5
APPROVED BY: JTC	ANGLES ± 0.5
DATE: 2-10	FILE NO 071
SCALE: 1:2	SHEET 2 OF 2

NAME

FILE SEC

GRADE 23

TANGENCIES

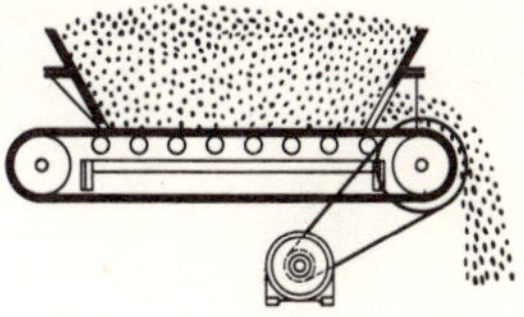

COURTESY JEFFREY MFG.

1 CONSTRUCT TWO LINES FROM POINT P TANGENT TO THE CIRCLE AT A. SHOW CONSTRUCTION AND LOCATE POINTS OF TANGENCY.

P

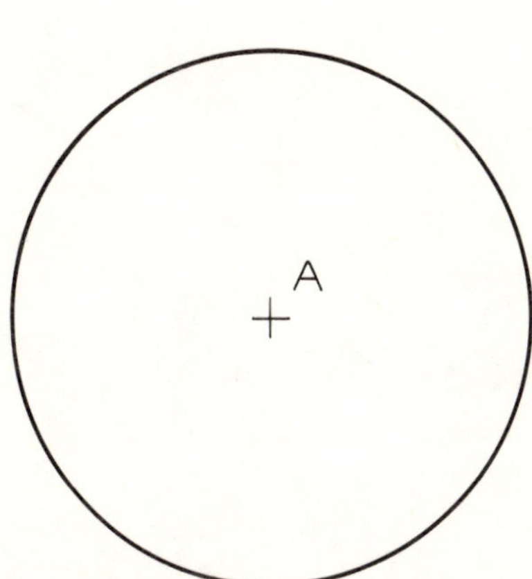

2 THE CONVEYOR SYSTEM ILLUSTRATES SEVERAL EXAMPLES OF TANGENCIES BETWEEN LINES AND CIRCLES. IN THE DRAWING BELOW, USE CONSTRUCTION LINES TO LOCATE ALL POINTS OF TANGENCY.

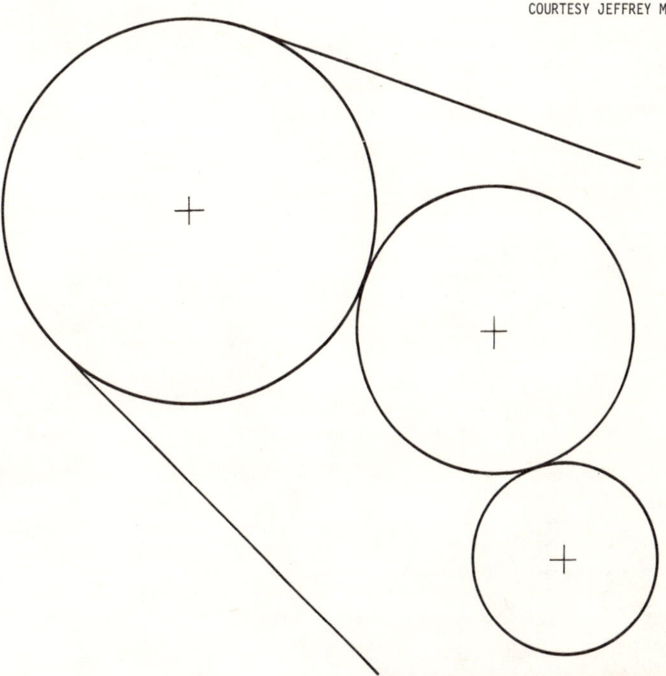

3 CONSTRUCT A 2" OR 50mm RADIUS ARC TANGENT TO THE TWO PERPENDICULAR LINES BELOW. SHOW CONSTRUCTION AND LOCATE POINTS OF TANGENCY.

4 COMPLETE THE BEND IN THE PIPE USING A CENTERLINE RADIUS OF 1.5" OR 40mm. SHOW CONSTRUCTION AND LOCATE POINTS OF TANGENCY. EXTEND LINES TO TANGENCY POINTS.

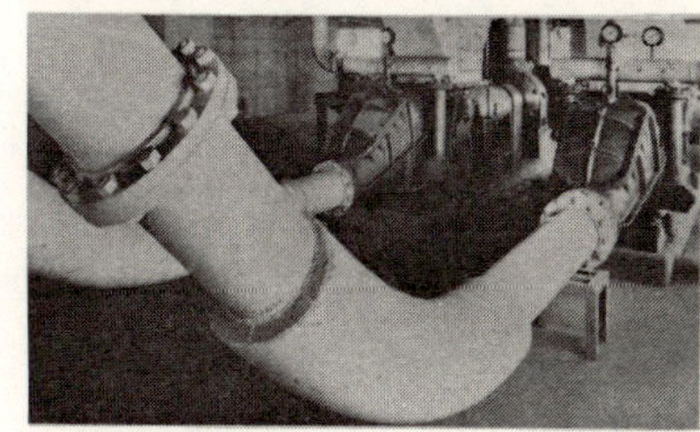

COURTESY FULLER CO.

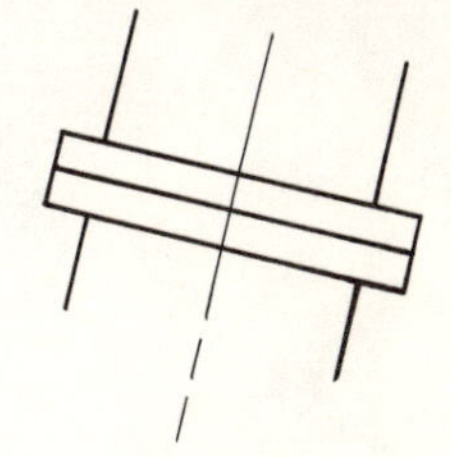

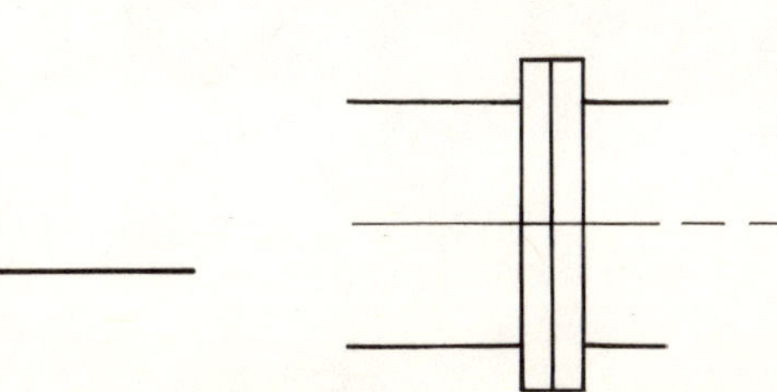

REB

TANGENCIES

1 COMPLETE THE RIGHT HALF OF THE WING HEAD BELOW USING A 2.5" OR 60mm RADIUS ARC TANGENT TO THE GIVEN ARC AND THE BASE LINE. SHOW CONSTRUCTION AND LOCATE POINTS OF TANGENCY.

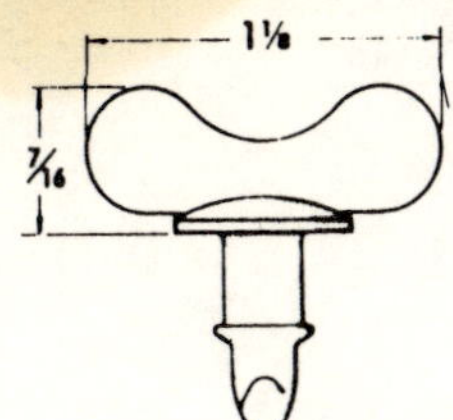

2 COMPLETE THE FILLET PORTION OF THE PILLOW BLOCK USING A 1" OR 25mm RADIUS TANGENT TO THE OUTER ARC AND TOP BASE LINE. SHOW CONSTRUCTION AND LOCATE POINTS OF TANGENCY.

COURTESY DODGE MFG.

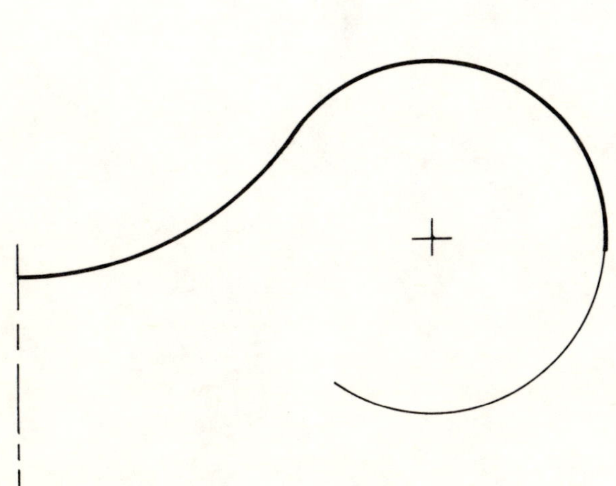

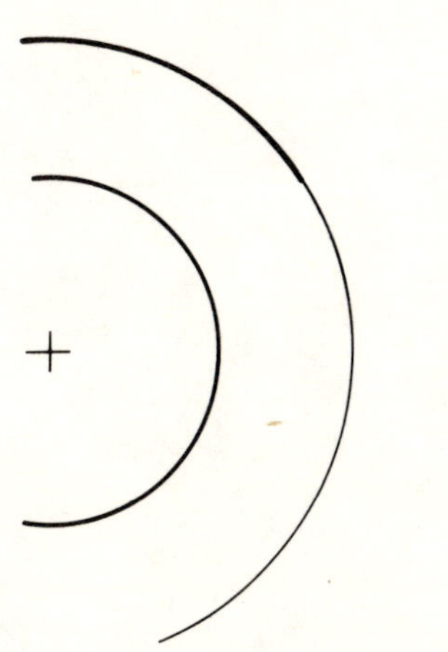

3 THE ROCKER PLATE BELOW SHOWS EXAMPLES OF BOTH CONVEX AND CONCAVE ARCS TANGENT TO TWO GIVEN ARCS. COMPLETE THE DRAWING AT THE LEFT USING A CONVEX ARC OF 4" OR 100mm RADIUS TANGENT ON THE TOP TO THE TWO LARGE ARCS, AND A CONCAVE ARC OF 1" OR 25mm RADIUS TANGENT ON THE BOTTOM. SHOW CONSTRUCTION AND LOCATE POINTS OF TANGENCY. EXTEND ARCS TO POINTS OF TANGENCY WHERE NECESSARY.

CONVEX

CONCAVE

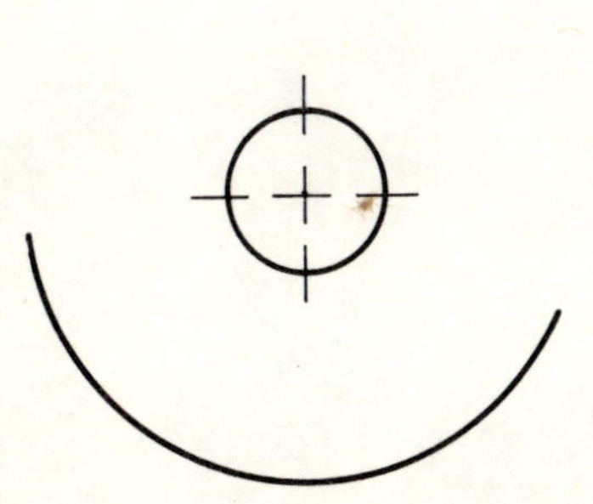

REB

SIX VIEW SKETCHING

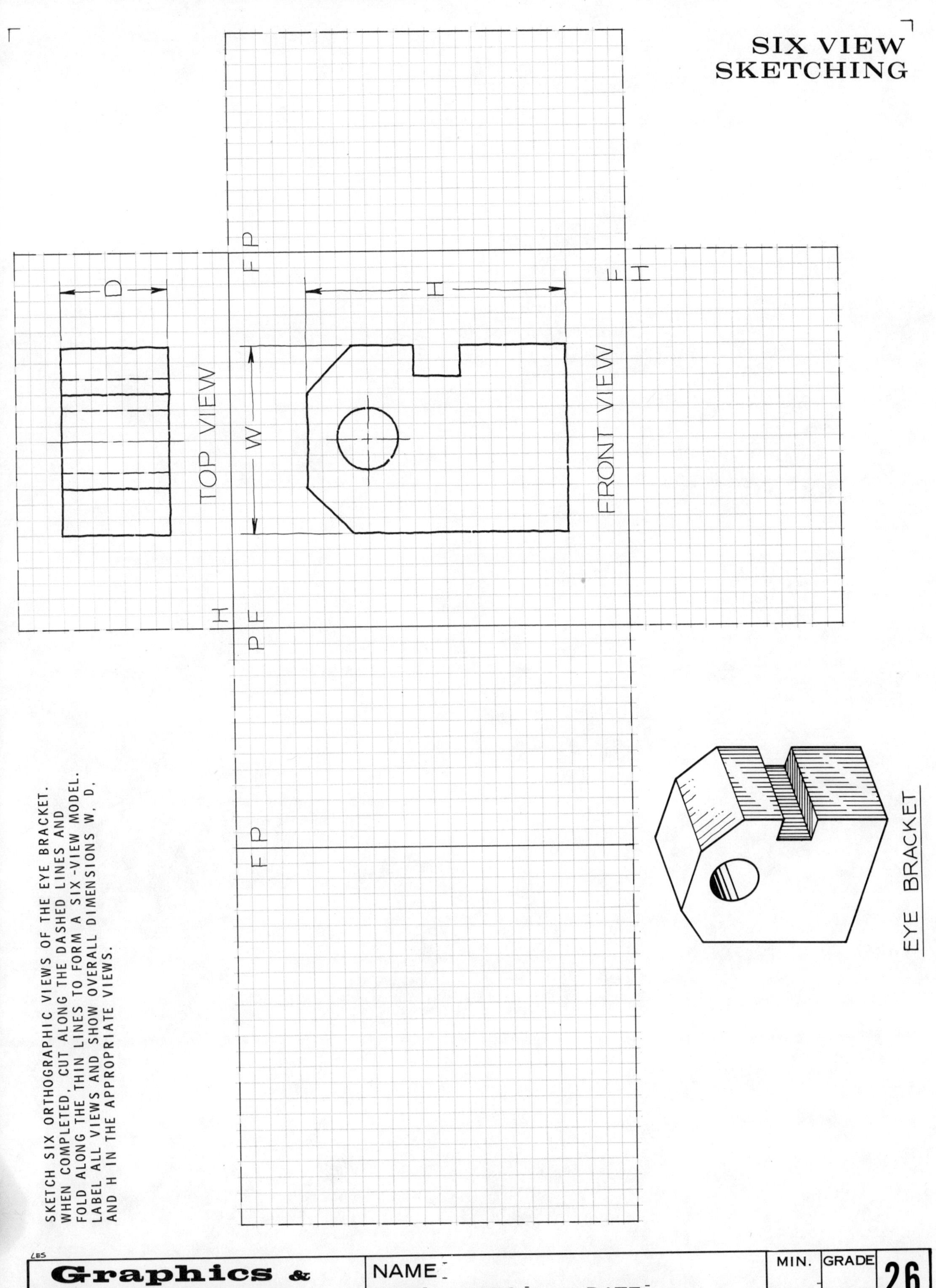

SKETCH SIX ORTHOGRAPHIC VIEWS OF THE EYE BRACKET. WHEN COMPLETED, CUT ALONG THE DASHED LINES AND FOLD ALONG THE THIN LINES TO FORM A SIX-VIEW MODEL. LABEL ALL VIEWS AND SHOW OVERALL DIMENSIONS W, D, AND H IN THE APPROPRIATE VIEWS.

THREE VIEW SKETCHING

1 SKETCH 3 VIEWS OF THE PARTS, LABEL THE VIEWS, AND SHOW DIMENSIONS W, D, AND H. PROBLEM 1: ONE PICTORIAL GRID IS EQUAL TO ONE ORTHOGRAPHIC GRID. PROBLEM 2: USE THE FULL SIZE DIMENSIONS GIVEN ON THE PICTORIAL.

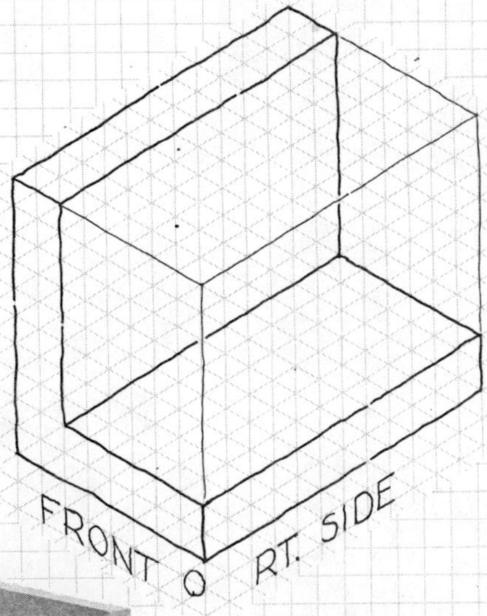

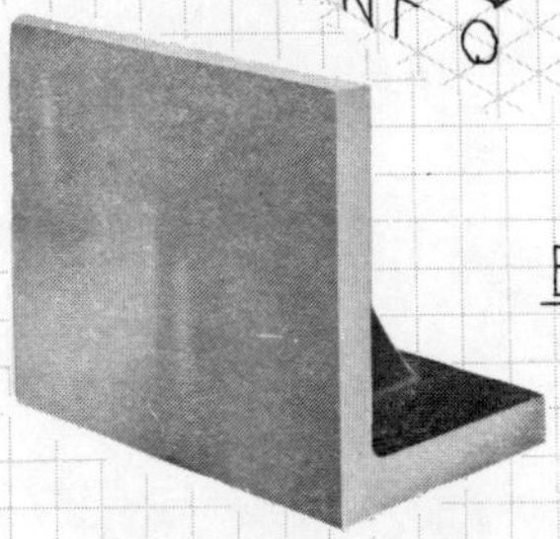

FRONT
RT. SIDE

ANGLE BRACKET

COURTESY THE CHALLENGE MACHINERY COMPANY

2

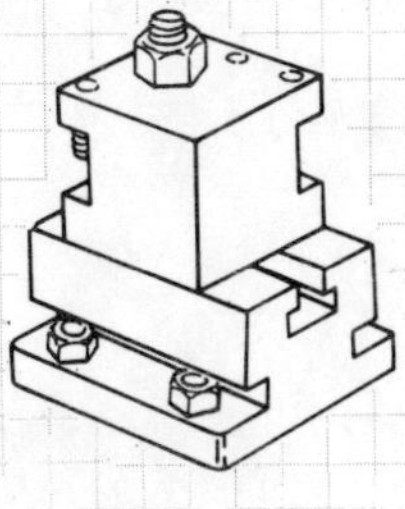

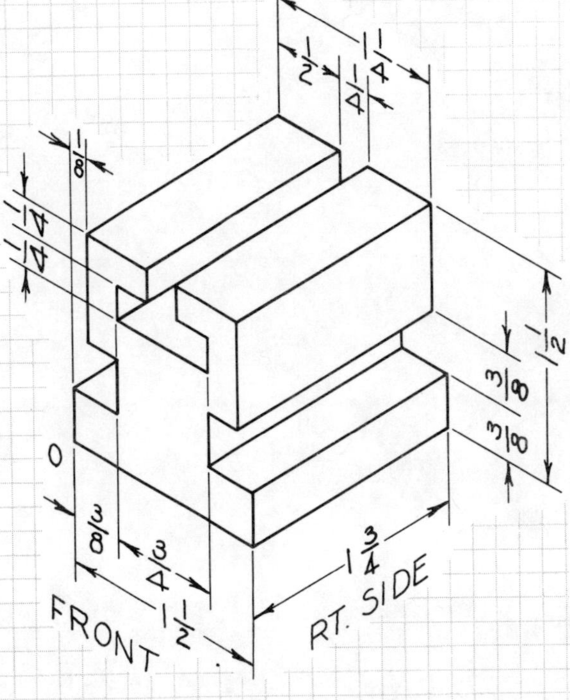

1/8
1/4
1/4
1/2
1/4
1 1/4
1 1/2
3/8
3/8
3/8
3/4
1 1/2
1 3/4
FRONT
RT. SIDE

Graphics & Geometry	NAME FILE SEC DATE	MIN.	GRADE	27

ORTHOGRAPHIC PROJECTION

SUPPLY THE MISSING LINES WITH INSTRUMENTS OR BY SKETCHING AS ASSIGNED BY YOUR INSTRUCTOR. SKETCH THE MORE DIFFICULT PROBLEMS PICTORIALLY ON THE GRID SHEETS IN THE BACK OF THIS PROBLEMS BOOK.

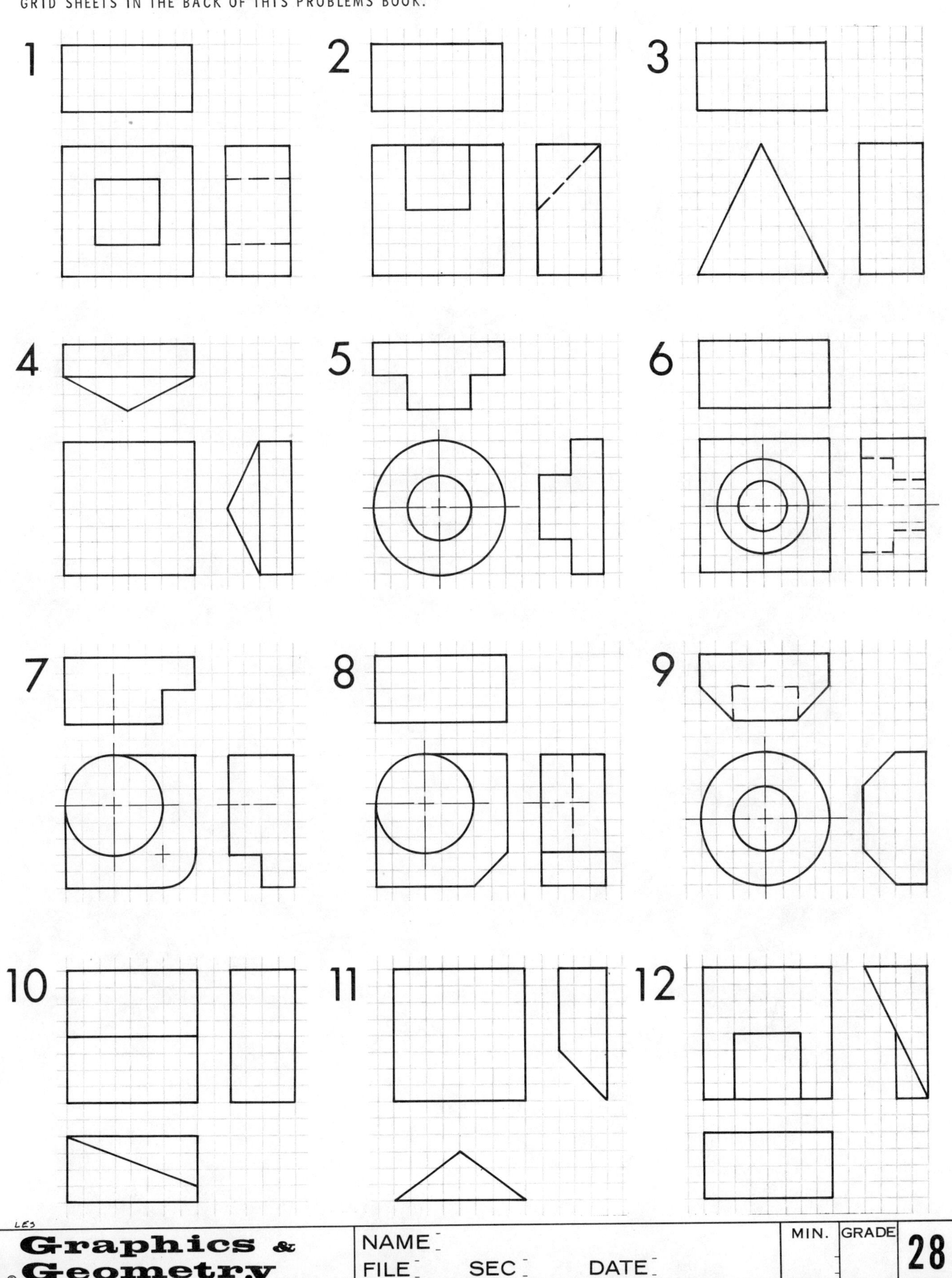

THE EYE BRACKET IS DESIGNED TO FASTEN A HYDRAULIC CYLINDER TO A MACHINERY SURFACE BY PIN AND CLEVIS.

REDESIGN THE VERTICAL PORTION OF THE EYE BRACKET BY MAKING IT SEMI-CYLINDRICAL WITH A RADIUS OF .75" OR 20mm CONCENTRIC TO THE .75" OR 20mm DIA HOLE. THE SIDES SHALL BE VERTICAL RATHER THAN ANGULAR. EACH CORNER OF THE BASE SHALL HAVE A RADIUS OF .38" OR 10mm CONCENTRIC WITH THE FOUR .38" OR 10mm DIAMETER HOLES.

MAKE A THREE-VIEW ORTHOGRAPHIC SKETCH OF THE PART THAT INCORPORATES THESE CHANGES.

FULL SIZE

DESIGN MODIFICATION

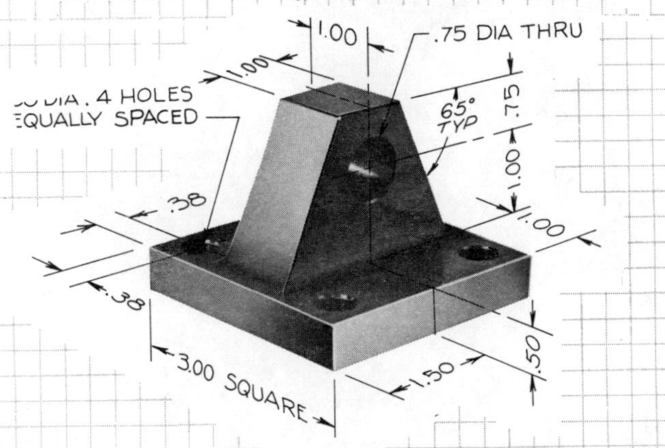

Courtesy of Dodge Manufacturing Co.

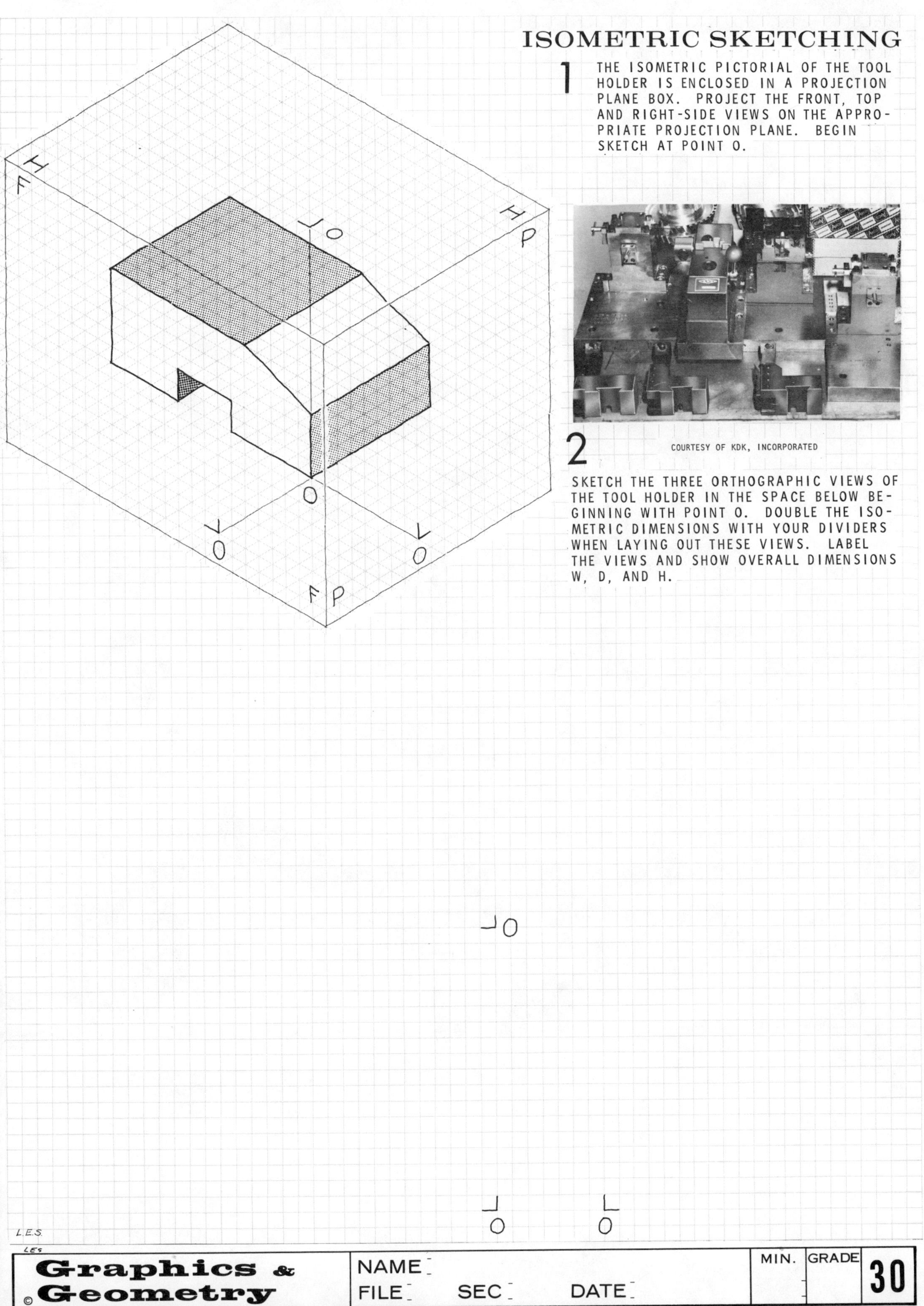

ISOMETRIC SKETCHING

1 THE ISOMETRIC PICTORIAL OF THE TOOL HOLDER IS ENCLOSED IN A PROJECTION PLANE BOX. PROJECT THE FRONT, TOP AND RIGHT-SIDE VIEWS ON THE APPROPRIATE PROJECTION PLANE. BEGIN SKETCH AT POINT O.

COURTESY OF KDK, INCORPORATED

2 SKETCH THE THREE ORTHOGRAPHIC VIEWS OF THE TOOL HOLDER IN THE SPACE BELOW BEGINNING WITH POINT O. DOUBLE THE ISOMETRIC DIMENSIONS WITH YOUR DIVIDERS WHEN LAYING OUT THESE VIEWS. LABEL THE VIEWS AND SHOW OVERALL DIMENSIONS W, D, AND H.

L.E.S.

Graphics & Geometry ©	NAME FILE SEC DATE	MIN.	GRADE	30

ISOMETRICS

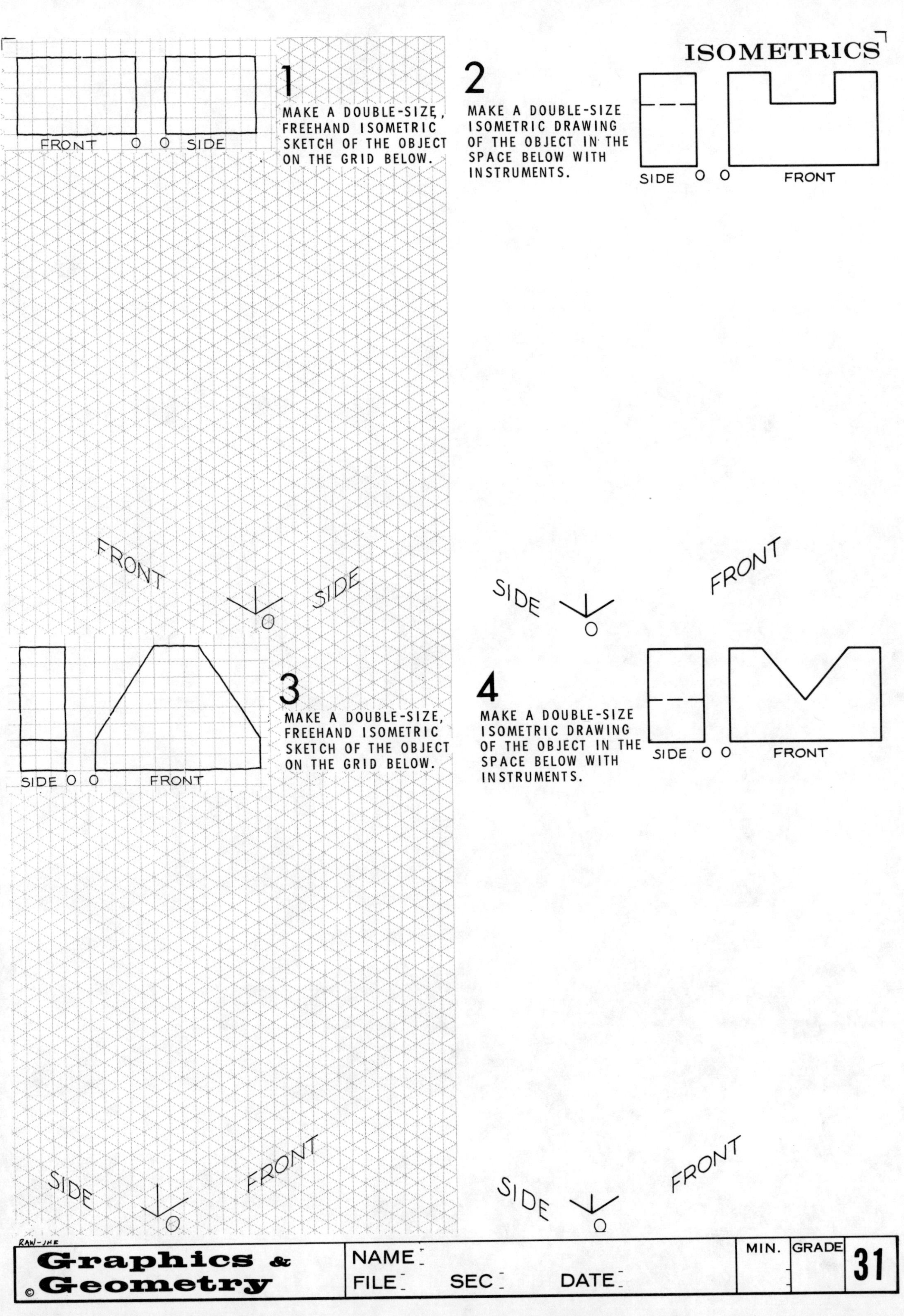

FRONT
SIDE
1
MAKE A DOUBLE-SIZE, FREEHAND ISOMETRIC SKETCH OF THE OBJECT ON THE GRID BELOW.
2
MAKE A DOUBLE-SIZE ISOMETRIC DRAWING OF THE OBJECT IN THE SPACE BELOW WITH INSTRUMENTS.
SIDE
FRONT
FRONT
SIDE
SIDE
FRONT
SIDE
FRONT
3
MAKE A DOUBLE-SIZE, FREEHAND ISOMETRIC SKETCH OF THE OBJECT ON THE GRID BELOW.
4
MAKE A DOUBLE-SIZE ISOMETRIC DRAWING OF THE OBJECT IN THE SPACE BELOW WITH INSTRUMENTS.
SIDE
FRONT
SIDE
FRONT
SIDE
FRONT

1 CONSTRUCT THREE ISOMETRIC CYLINDERS, USING YOUR ISOMETRIC TEMPLATE AND THE FOLLOWING SPECIFICATIONS: (A) 1.0" or 25.5 mm DIA, (B) 1.38" or 35 mm DIA, (C) 1.5" or 38 mm DIA. THE END MARKED "V" IS VISIBLE.

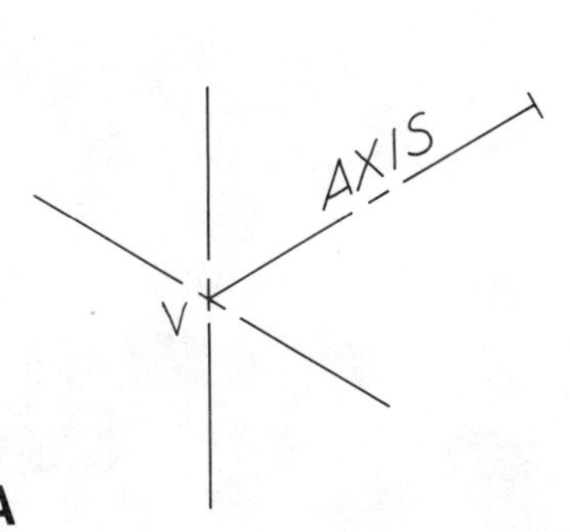

A

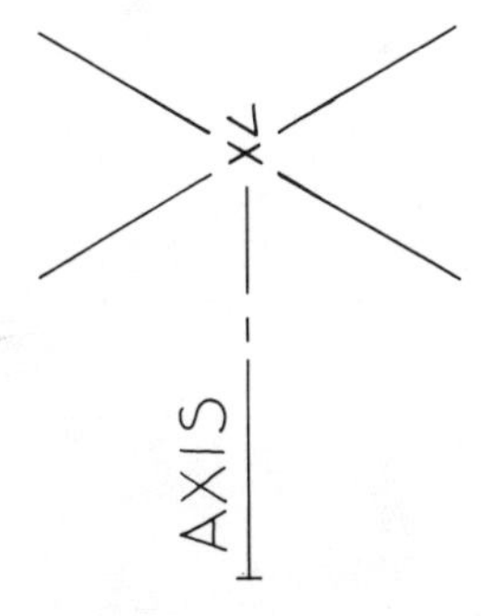

B

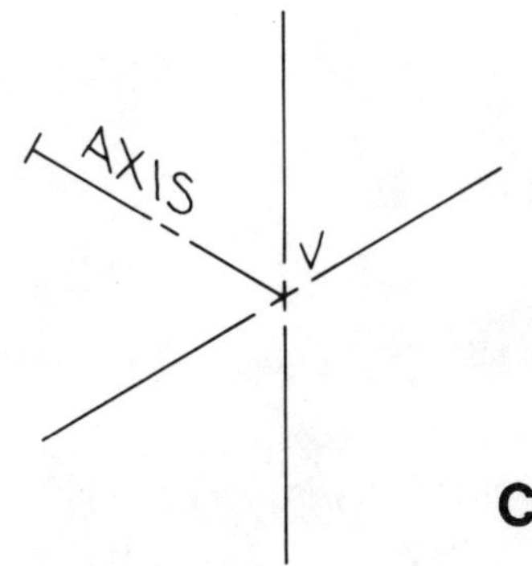

C

2 COMPLETE THE FULL-SIZE, PARTIAL ISOMETRIC DRAWING USING THE NOTED DIMENSIONS.

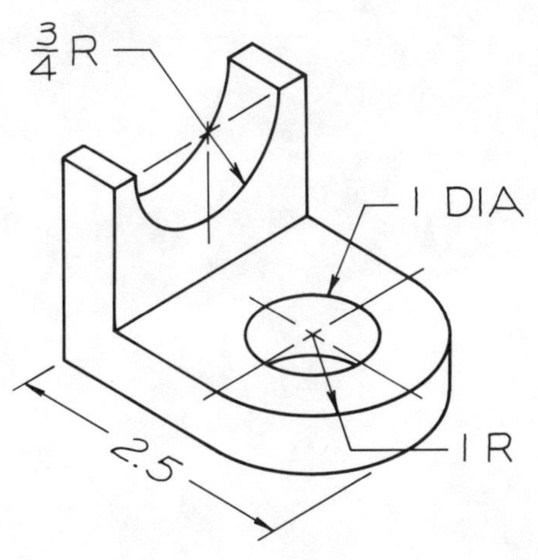

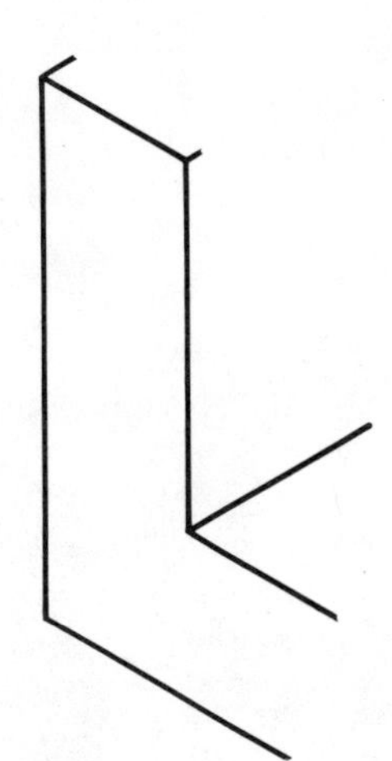

3 MAKE AN ISOMETRIC DRAWING OF A CIRCLE WITH A 3.5" (90 mm) DIAMETER. USE THE FOUR-CENTER METHOD WITH COMPASS.

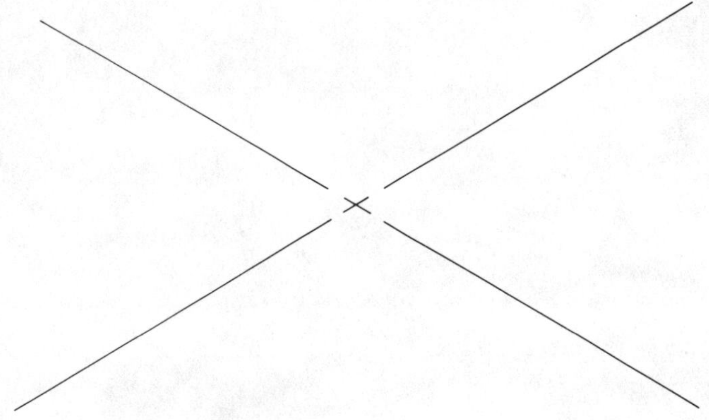

COURTESY BROWN & **ISOMETRICS**

FRONT SIDE O

FRONT SIDE O

1 MAKE A FULL SIZE ISOMETRIC DRAWING OF THE FITTING SHOWN BELOW. BEGIN AT "O".

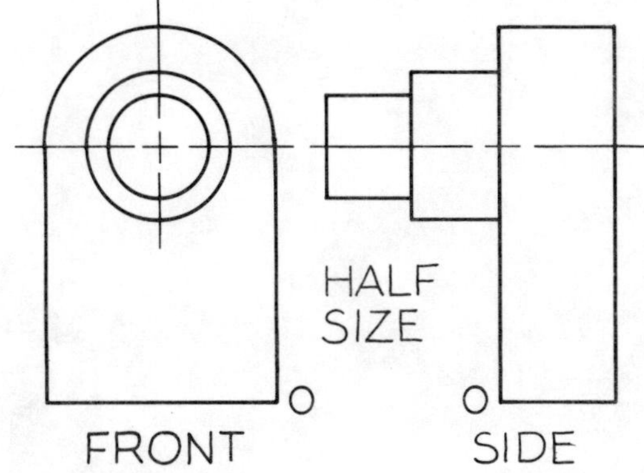

2 MAKE A FULL SIZE ISOMETRIC DRAWING OF THE BEARING SUPPORT SHOWN BELOW.

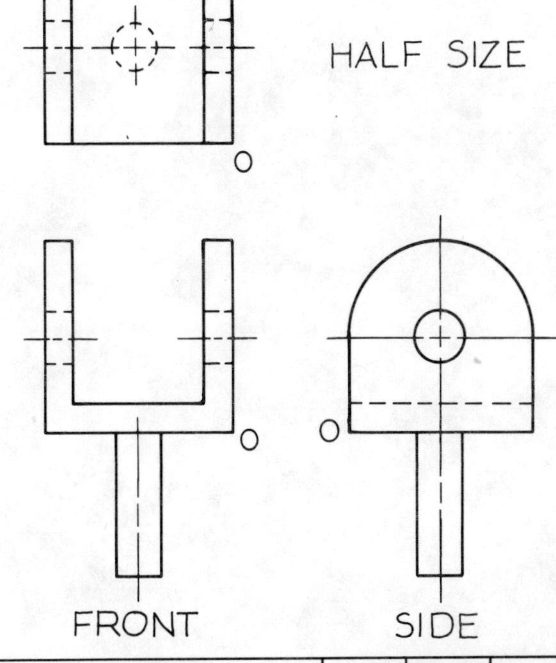

RW

OBLIQUES

1 USING INSTRUMENTS MAKE A CAVALIER AND CABINET DRAWING OF THE CHANNEL SHOWN. USE THE RECEDING AXES INDICATED AND DOUBLE THE GIVEN DIMENSIONS WITH DIVIDERS.

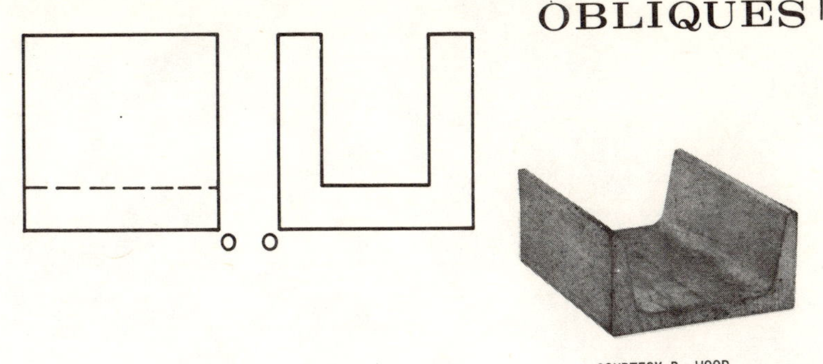

COURTESY B. WOOD

CAVALIER

CABINET

2 WITH INSTRUMENTS MAKE A CAVALIER OBLIQUE DRAWING OF THE "T" NUT SHOWN. USE THE RECEDING AXES INDICATED AND DOUBLE THE GIVEN DIMENSIONS WITH DIVIDERS.

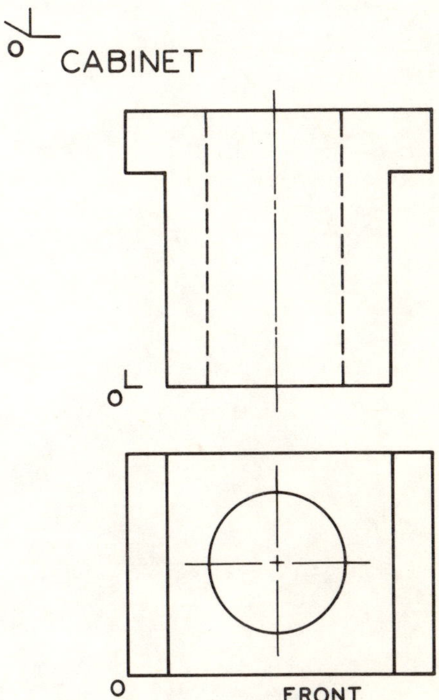

COURTESY CARR LANE

FRONT

G R

OBLIQUES

1 CONSTRUCT THE FULL SIZE CAVALIER AND CABINET DRAWINGS OF THE VERNIER BLANK. USE A 30° RECEDING AXIS. THE DIMENSIONS MAY BE DOUBLED BY USING DIVIDERS.

SCALE: FULL

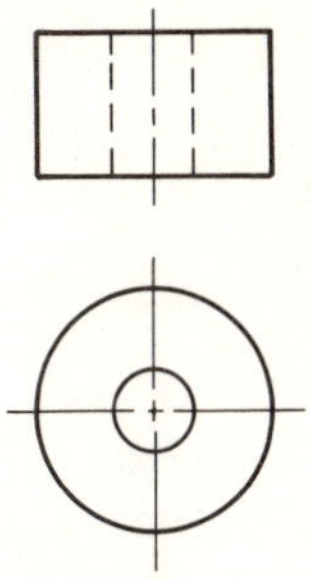

SCALE: HALF SIZE

CAVALIER

CABINET

O

O

2 CONSTRUCT THE FULL SIZE CAVALIER DRAWINGS OF THE JAW. BEGIN AT POINT O AND DOUBLE THE DIMENSIONS USING DIVIDERS.

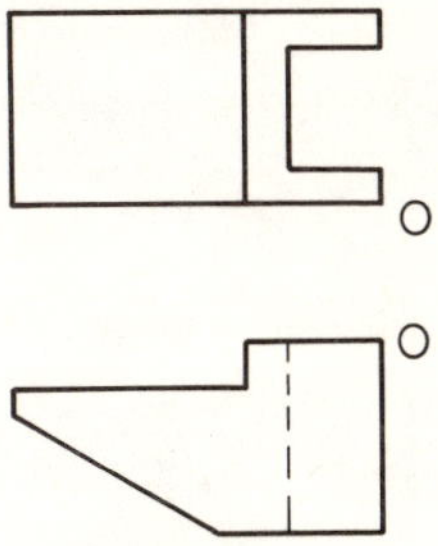

O

O

SCALE: HALF SIZE

SCALE: FULL

3 CONSTRUCT THE FULL SIZE CABINET OBLIQUE DRAWING OF THE BASE. BEGIN AT POINT P AND DOUBLE THE DIMENSIONS USING DIVIDERS.

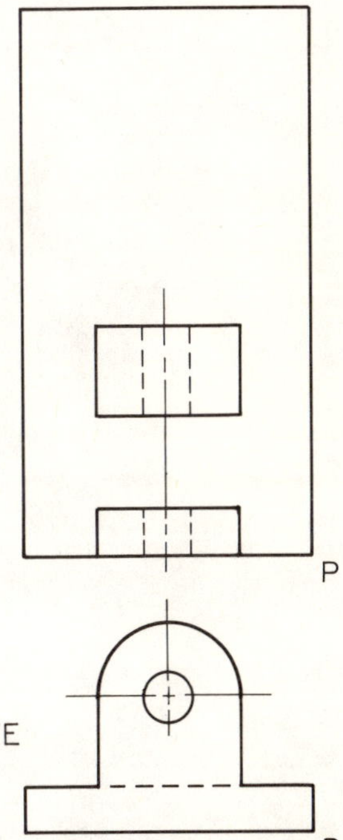

P

SCALE: HALF SIZE

P

FRONT

FRONT

P

SCALE: FULL

JTC

Graphics & Geometry	NAME FILE SEC DATE	MIN.	GRADE	36

ENGLISH UNITS

DRAW THE MISSING VIEWS OF THE PARTS BELOW. SEE THAT ALL VIEWS ARE CORRECT.

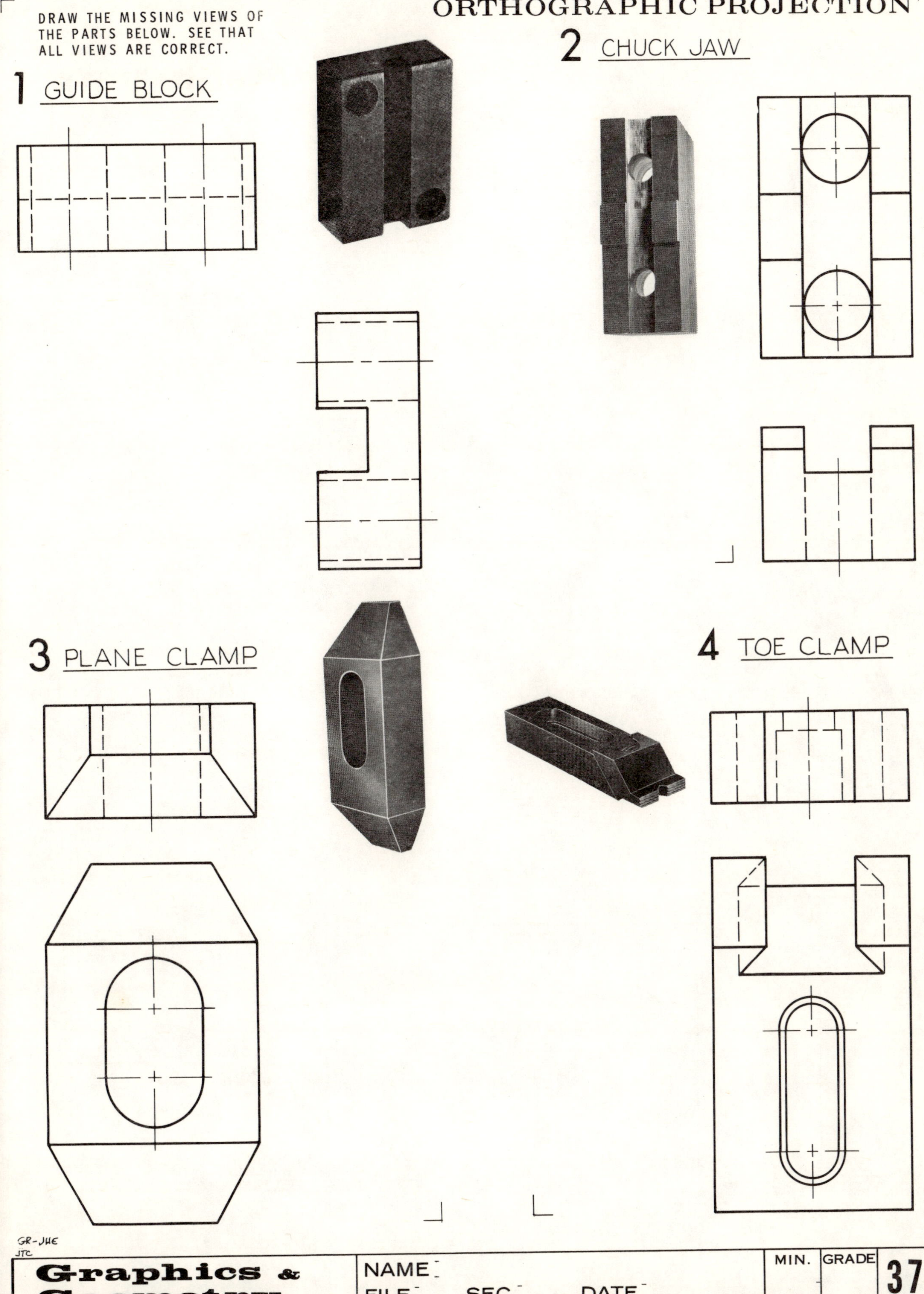

GR-JHE
JTC

ENGLISH UNITS

ENGINEERS FREQUENTLY MAKE ROUGH SKETCHES TO CONVEY IDEAS TO DRAFTSMEN. FREEHAND SKETCHES OF TWO PARTS ARE GIVEN BELOW WITH DIMENSIONS GIVEN IN MILLIMETERS.

ASSUME THAT YOU ARE THE DRAFTSMAN AND AN ENGINEER HAS GIVEN YOU THESE SKETCHES TO CONVERT INTO AN INSTRUMENT DRAWING. DRAW THE VIEWS IN THE SPACE BELOW AND STRIVE FOR THE PROPER CONTRAST BETWEEN THE TYPES OF LINES. OMIT DIMENSIONS. CORRECT ALL ERRORS, IF ANY.

ORTHOGRAPHIC PROJECTION

COURTESY OF BREWER MACHINE & GEAR CO.

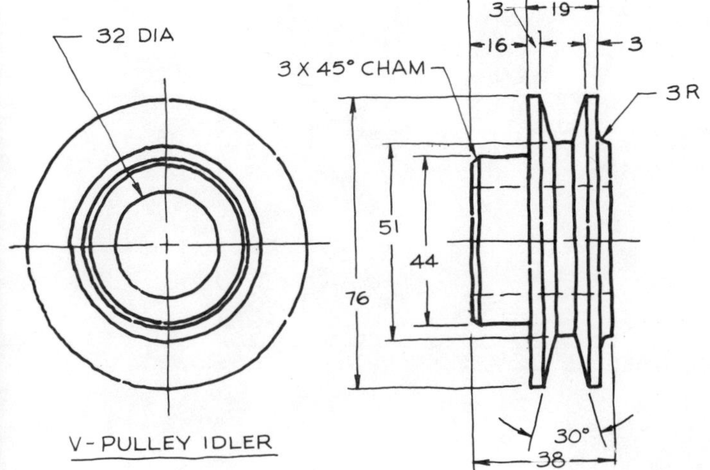

V-PULLEY IDLER

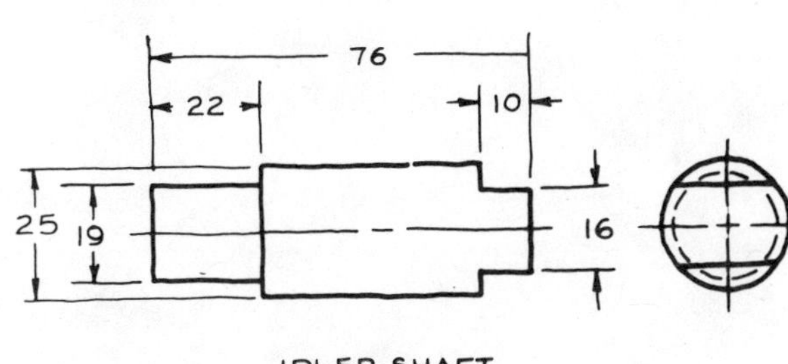

IDLER SHAFT

IDLER SHAFT

V-PULLEY IDLER

SI

FULL SIZE

ENGLISH UNITS

PIPING DRAWING

ORTHOGRAPHIC AND ISOMETRIC DRAWINGS ARE COMMONLY USED TO REPRESENT PIPING SYSTEMS. SUCH AS SHOWN IN THE EXAMPLE BELOW.

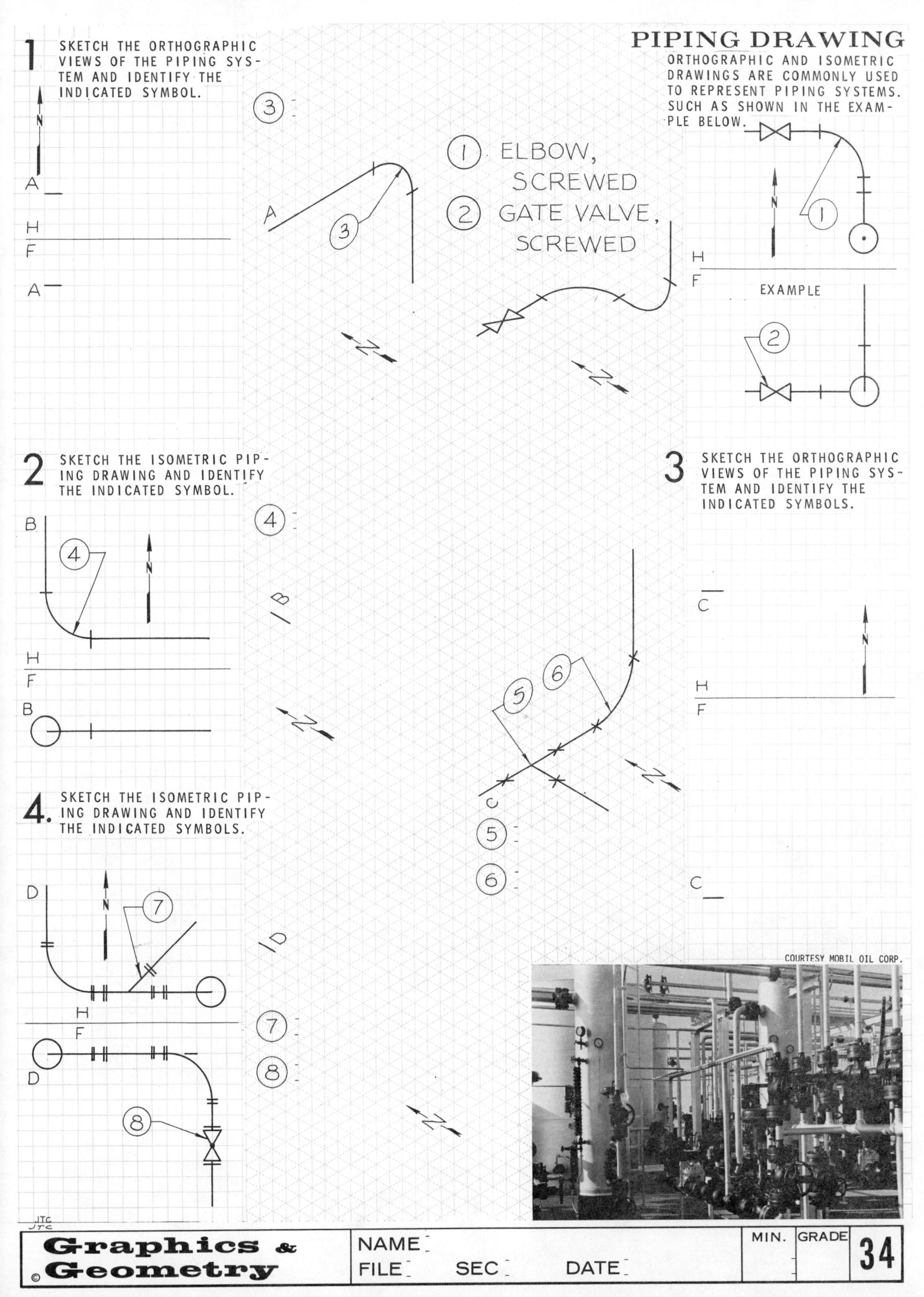

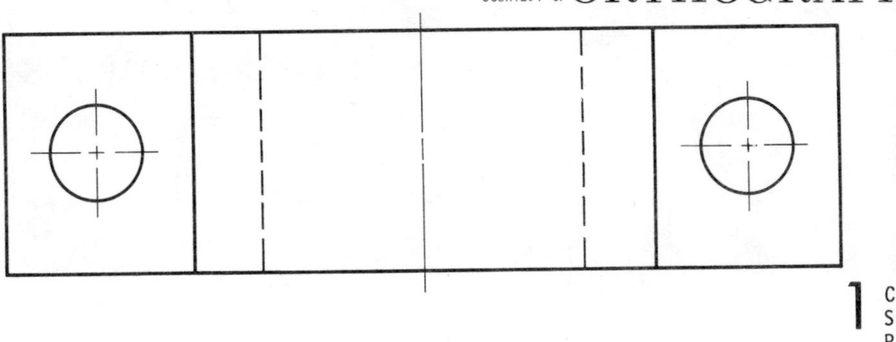

1 CONSTRUCT THE FRONT VIEW OF THE SHAFT BRACKET ACCORDING TO THE PHOTOGRAPH.

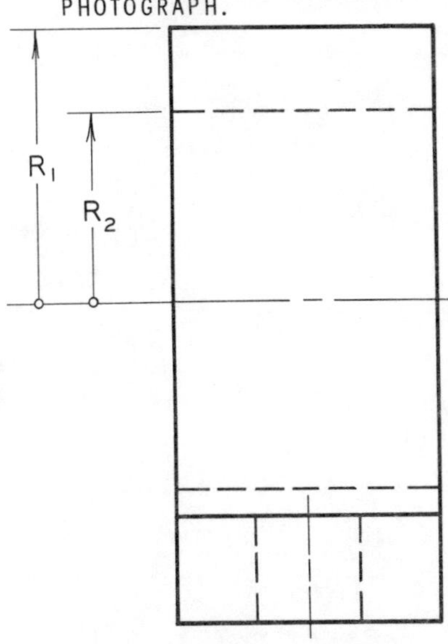

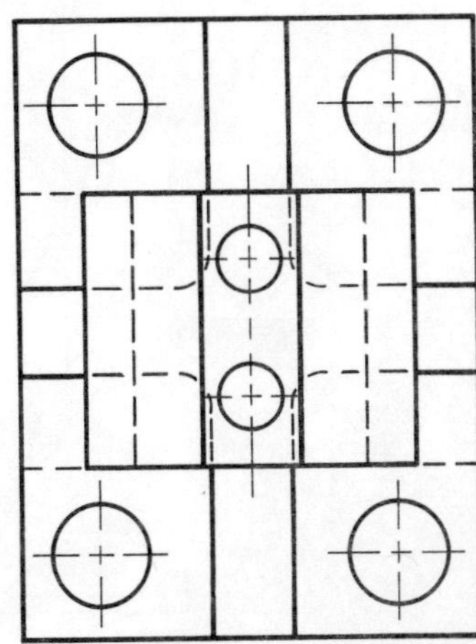

2 DRAW THE MISSING VIEW OF THE SUPPORT STAND.

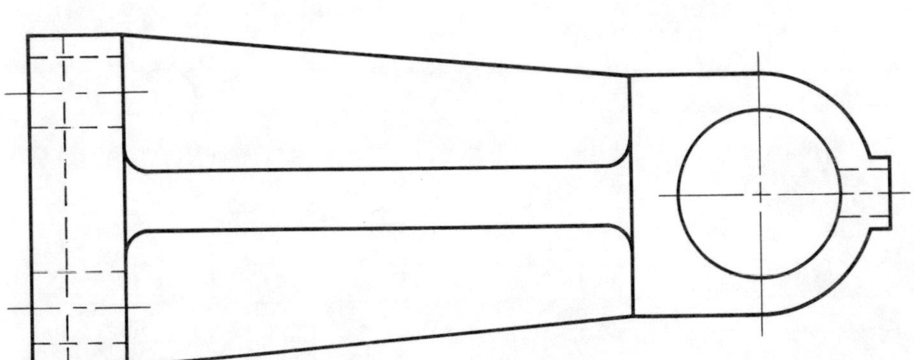

COURTESY OF THE JOHNSON CORPORATION

PMM

Graphics & Geometry ©	NAME FILE SEC DATE	MIN.	GRADE	39

ENGLISH UNITS

ORTHOGRAPHIC PROJECTION

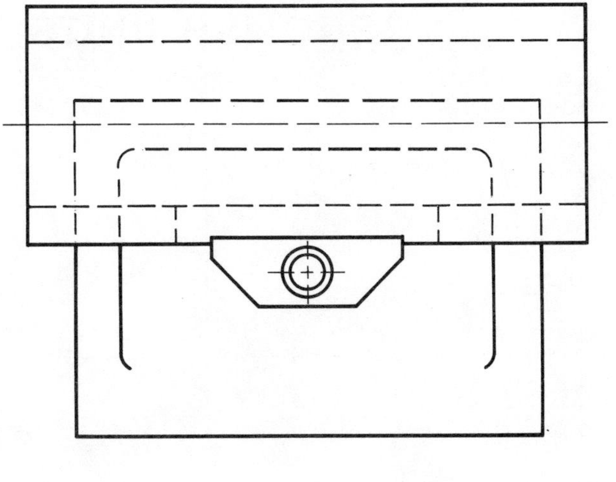

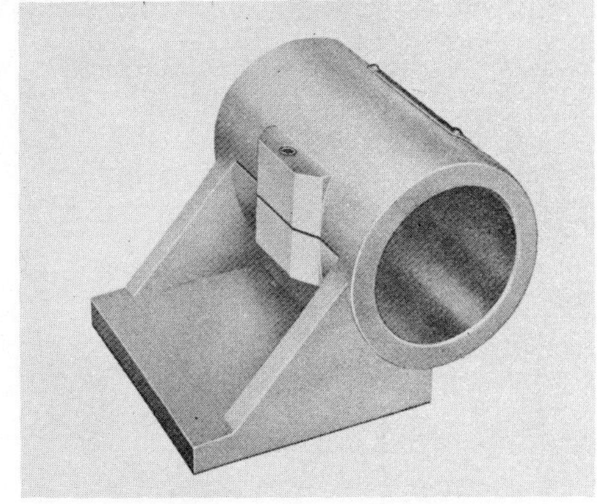

1 DRAW THE FRONT VIEW OF THE MOUNTING BRACKET.

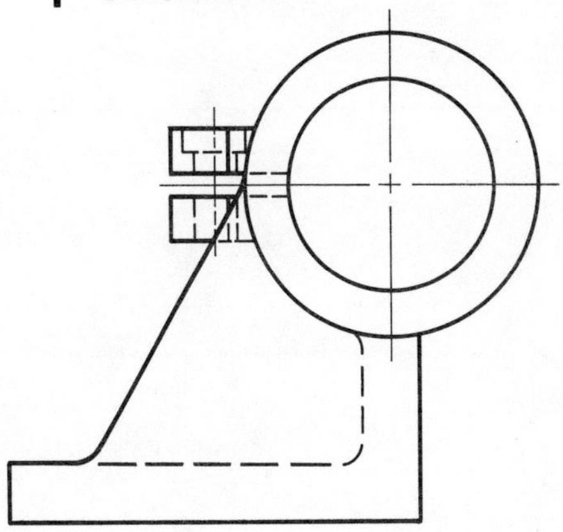

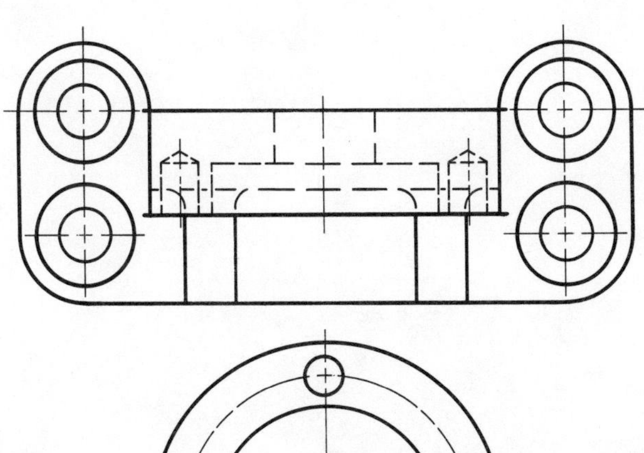

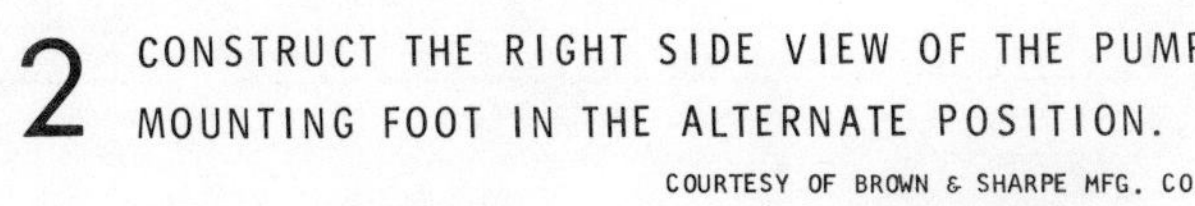

2 CONSTRUCT THE RIGHT SIDE VIEW OF THE PUMP MOUNTING FOOT IN THE ALTERNATE POSITION.

COURTESY OF BROWN & SHARPE MFG. CO.

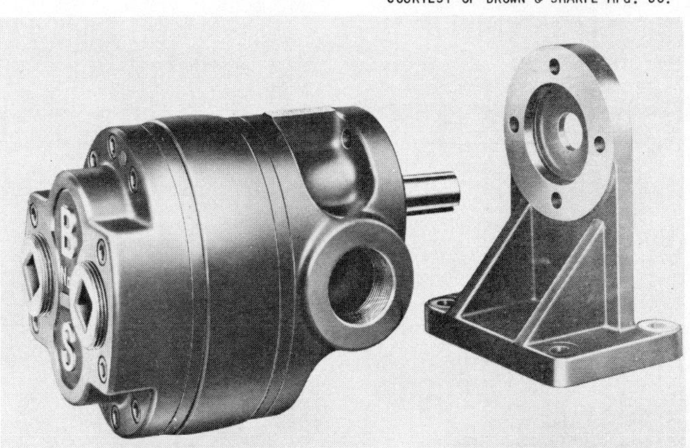

PMM

Graphics & Geometry ©	NAME FILE SEC DATE	MIN.	GRADE	40

ENGLISH UNITS

1 PROJECT THE FRONT, TOP AND PARTIAL RIGHT SIDE VIEWS ON THE PRINCIPAL AND AUXILIARY PLANES OF PROJECTION. POINT T IS PROJECTED TO ALL PLANES OF PROJECTION IN ORDER TO POSITION THE VIEWS.

2 SKETCH THE AUXILIARY VIEW IN THE SPACE PROVIDED BELOW. CUT OUT THE GRID AND FOLD ON THE REFERENCE PLANE LINES TO PRODUCE A THREE-DIMENSIONAL MODEL.

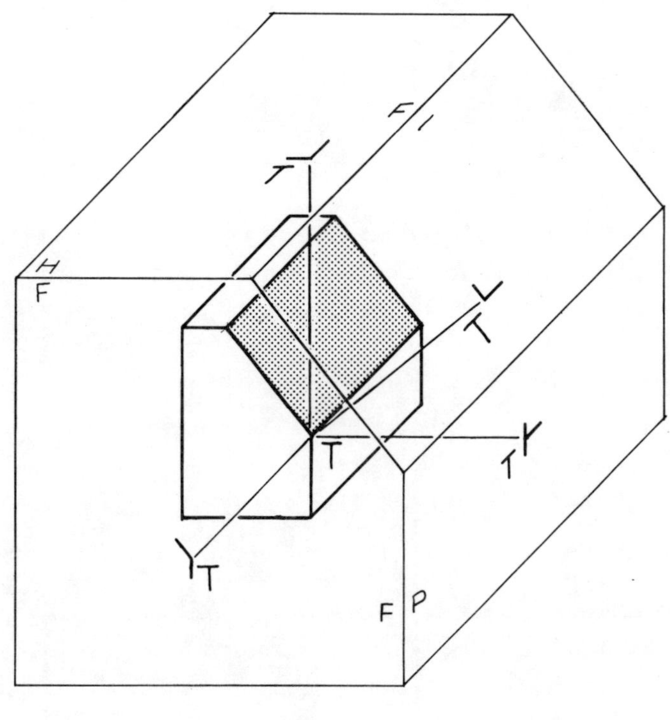

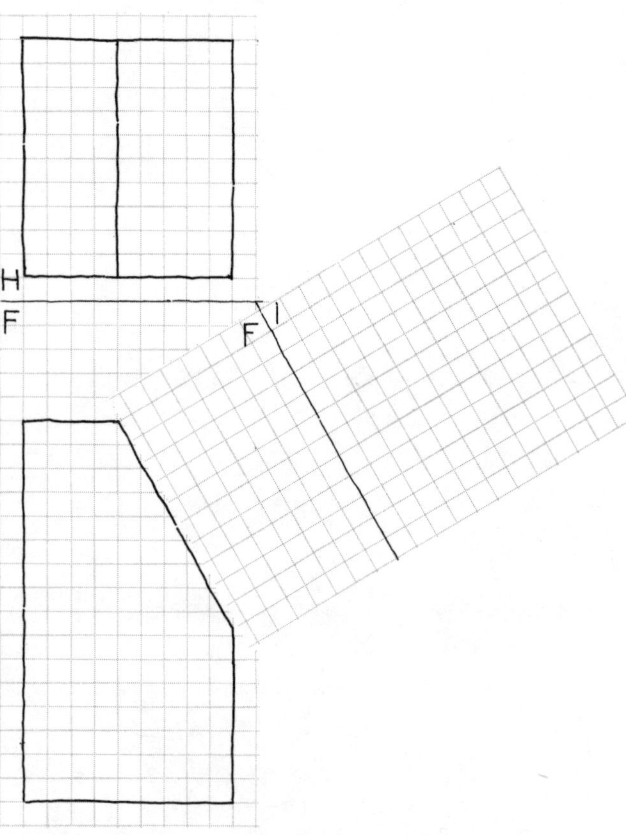

3 PROJECT THE PARTIAL FRONT, TOP AND AUXILIARY VIEWS ON THE PRINCIPAL PLANES OF PROJECTION. POINT P IS PROJECTED ON ALL PLANES OF PROJECTION TO POSITION THE VIEWS.

4 SKETCH THE AUXILIARY VIEW IN THE SPACE PROVIDED. CUT OUT THE GRID AREA. FOLD ON THE REFERENCE PLANE LINES TO PRODUCE A THREE-DIMENSIONAL MODEL.

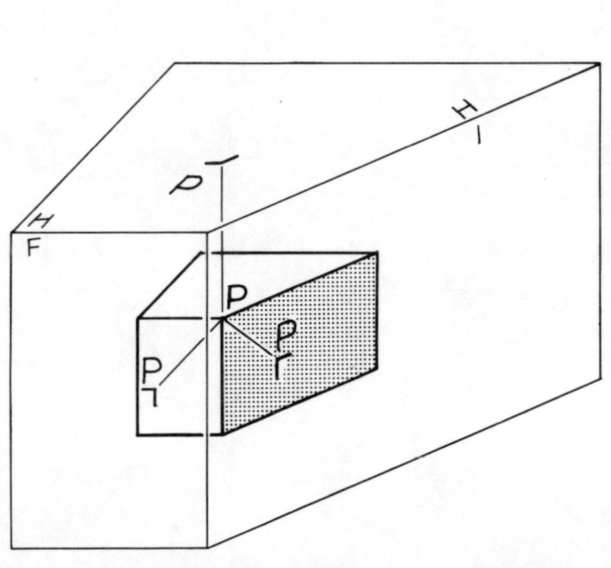

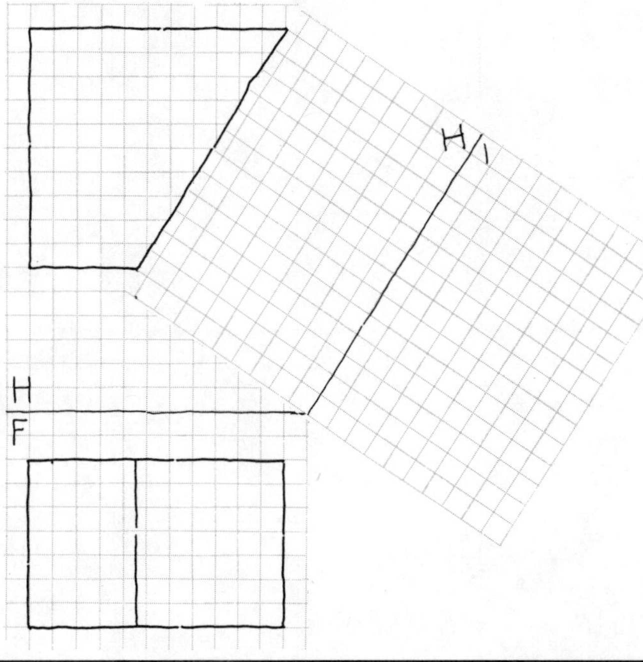

ENGLISH UNITS

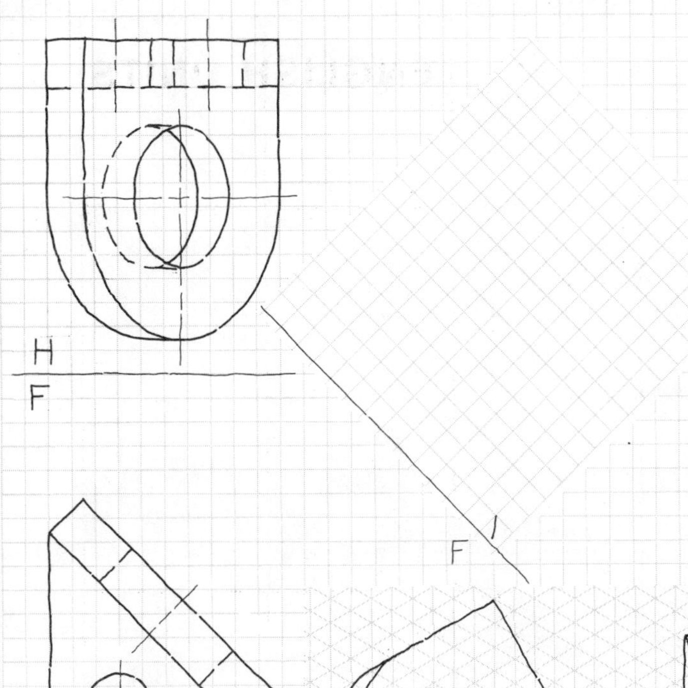

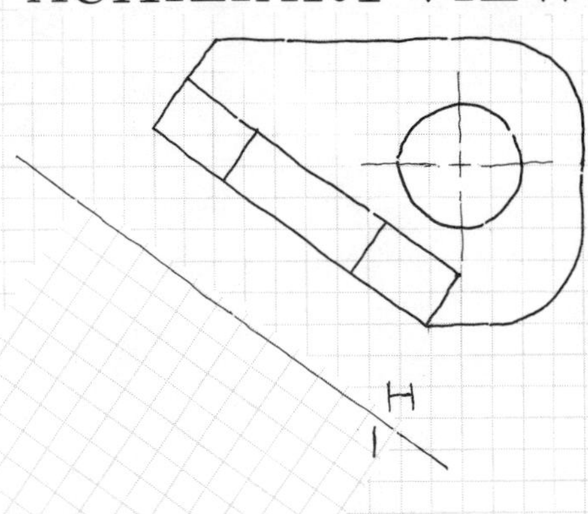

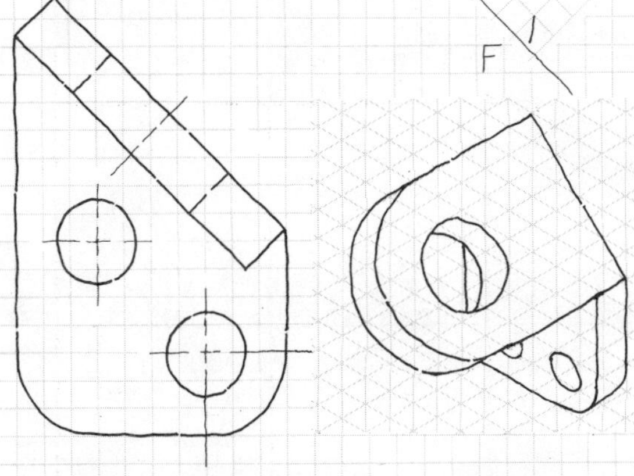

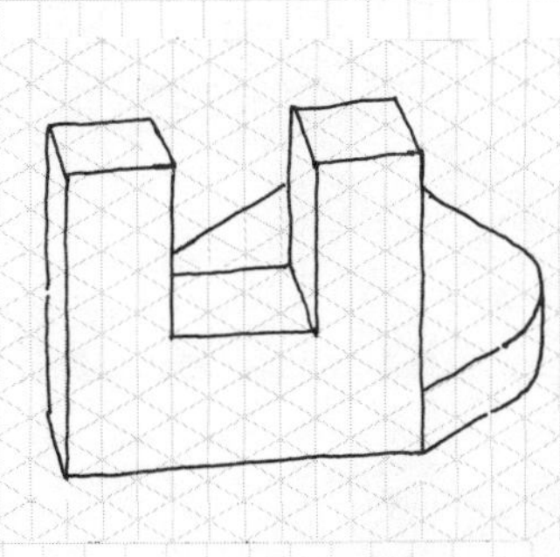

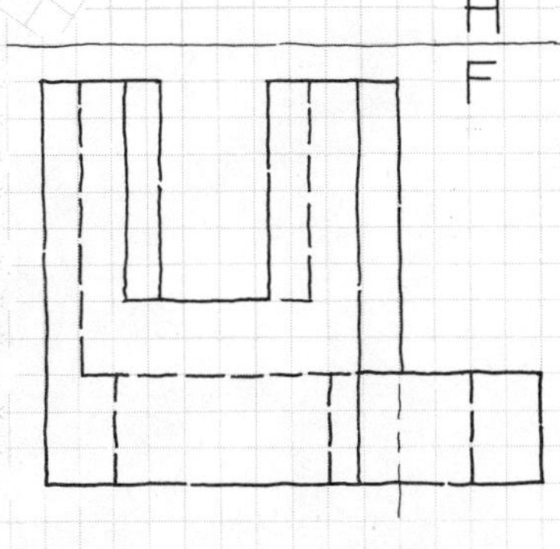

1 SKETCH AN AUXILIARY VIEW OF THE INCLINED SURFACE OF THE ANGLE SUPPORT SHOWN ABOVE.

2 DRAW AN AUXILIARY VIEW OF THE INCLINED SURFACE OF THE SWIVEL GUIDE SHOWN ABOVE.

3 DRAW AN AUXILIARY VIEW OF THE INCLINED SURFACE OF THE ANGLE GIB SHOWN BELOW.

4 DRAW AN AUXILIARY VIEW OF THE INCLINED SURFACE OF THE LATCH HOOK SHOWN BELOW.

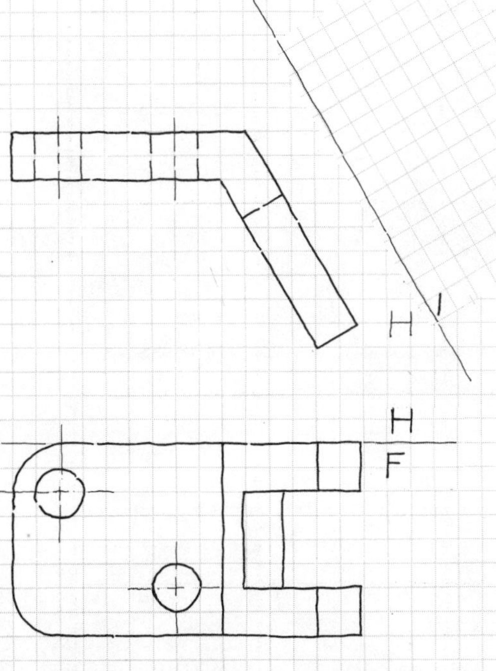

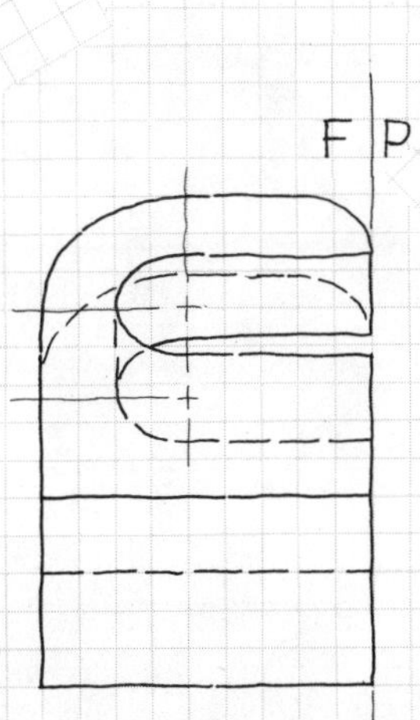

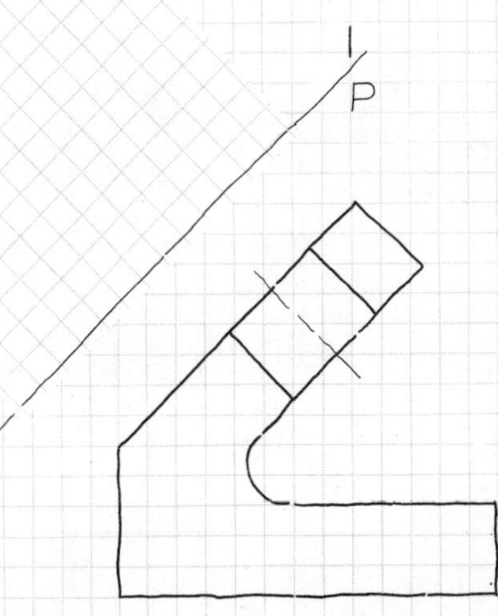

ENGLISH UNITS

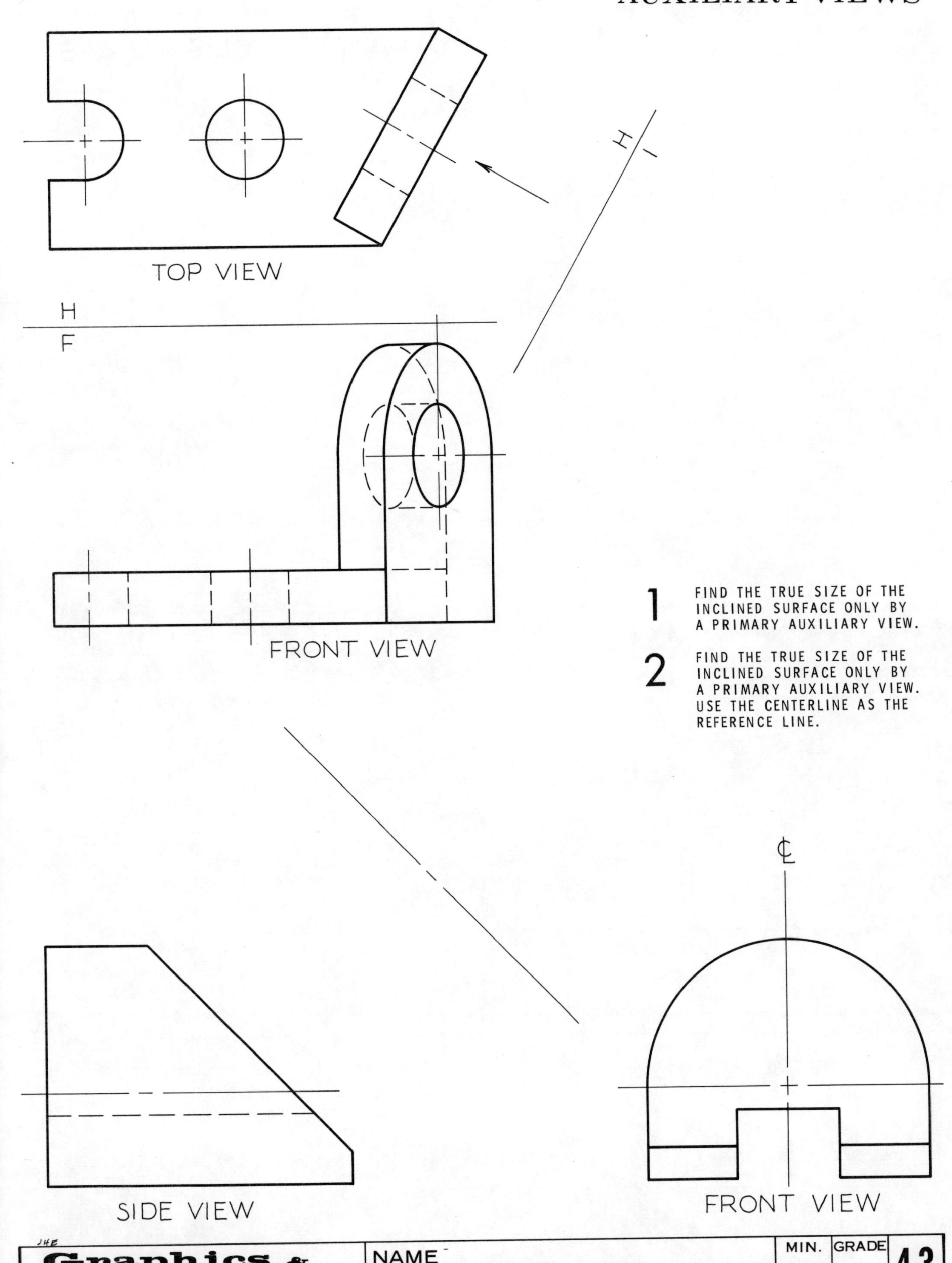

1 FIND THE TRUE SIZE OF THE INCLINED SURFACE ONLY BY A PRIMARY AUXILIARY VIEW.

2 FIND THE TRUE SIZE OF THE INCLINED SURFACE ONLY BY A PRIMARY AUXILIARY VIEW. USE THE CENTERLINE AS THE REFERENCE LINE.

ENGLISH UNITS

FULL SECTIONS

1 COMPLETE THE FRONT VIEW AS A FULL SECTION. USE CAST IRON SECTION SYMBOLS.

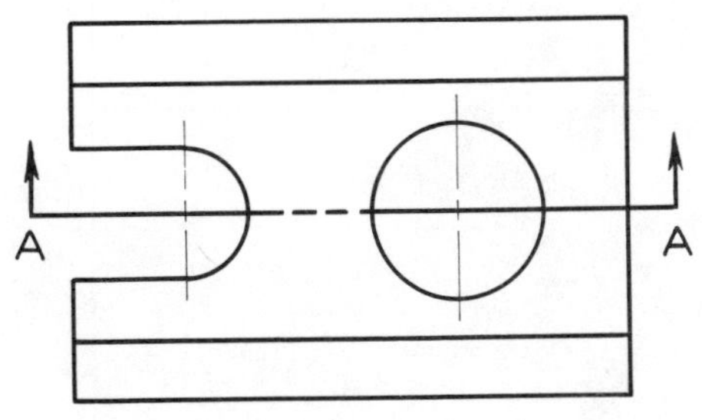

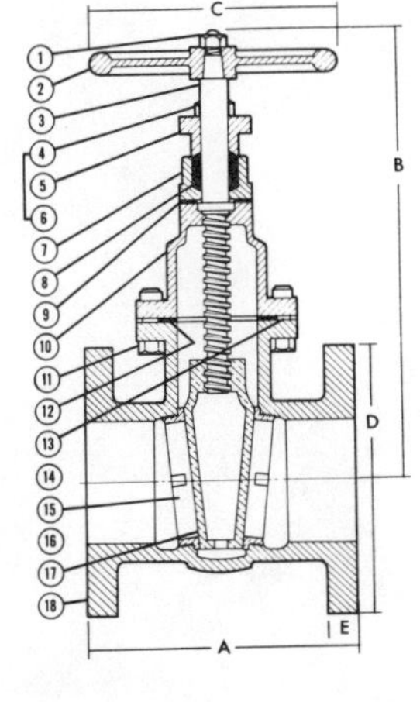

2 COMPLETE THE FRONT VIEW AS A FULL SECTION. SHOW CUTTING PLANE AND USE STEEL SYMBOLS.

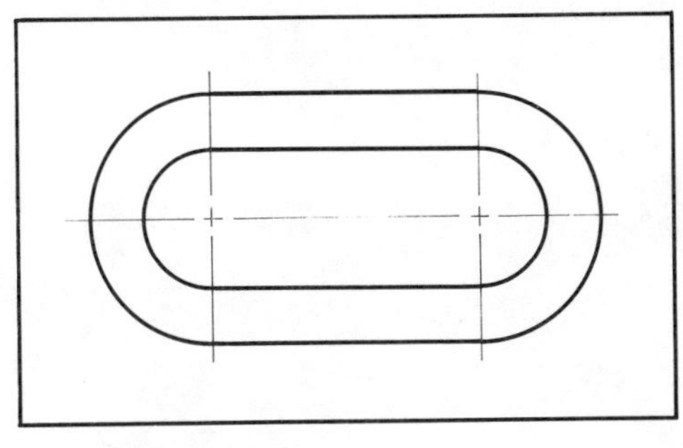

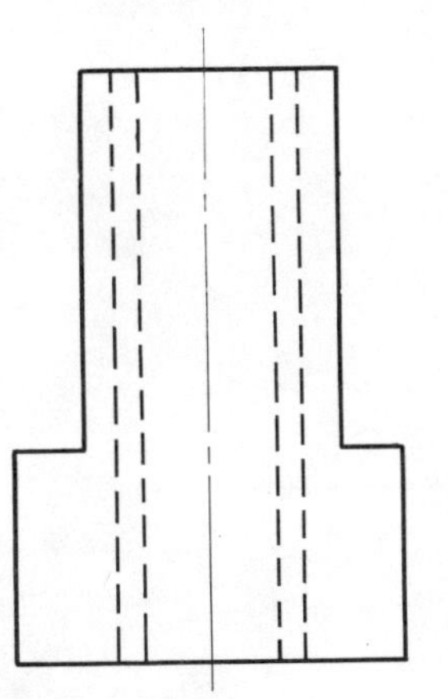

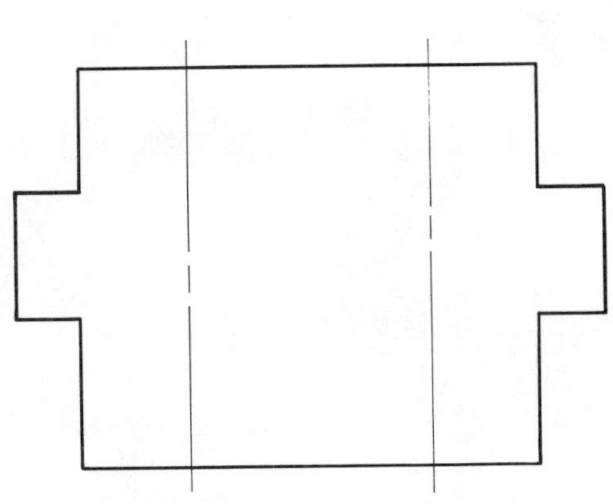

SECTION A-A

3 COMPLETE THE PARTIAL FRONT VIEW AS A FULL SECTION. SHOW CUTTING PLANE. USE BRASS SYMBOLS.

4 COMPLETE THE FRONT VIEW AS A FULL SECTION. USE CAST IRON SYMBOLS. CUTTING PLANE UNNECESSARY.

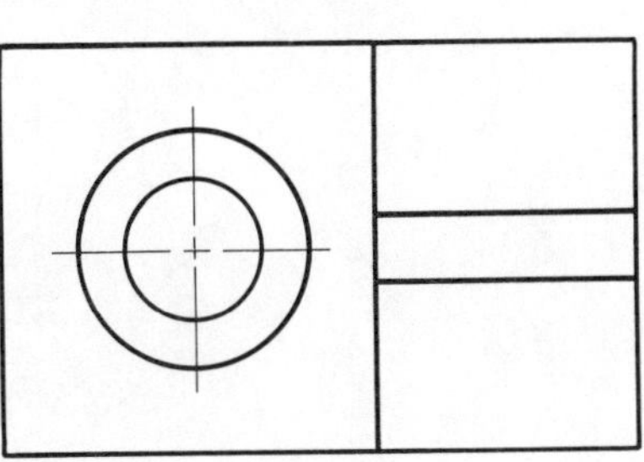

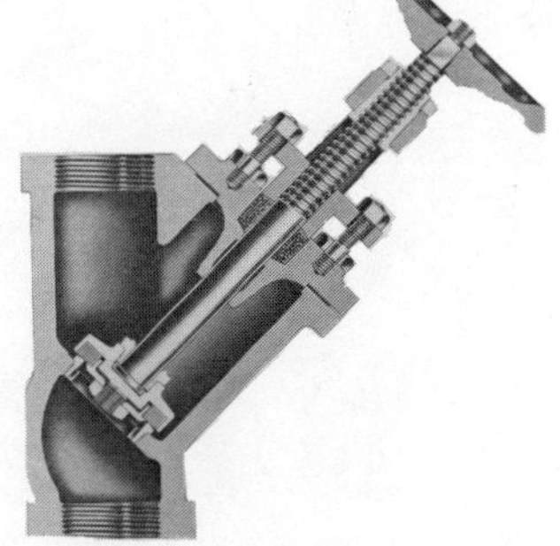

COURTESY OF JENKINS BROS.

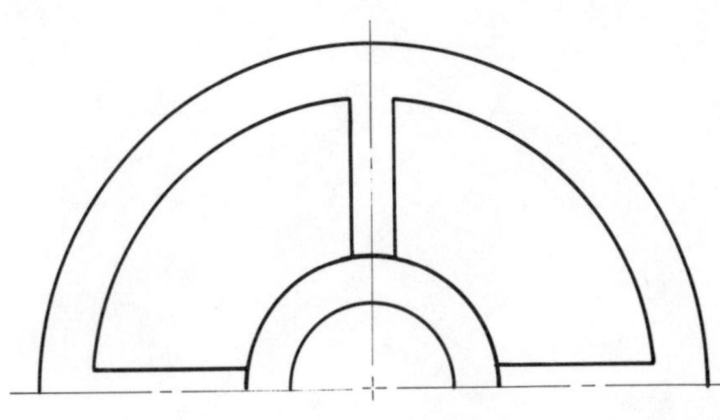

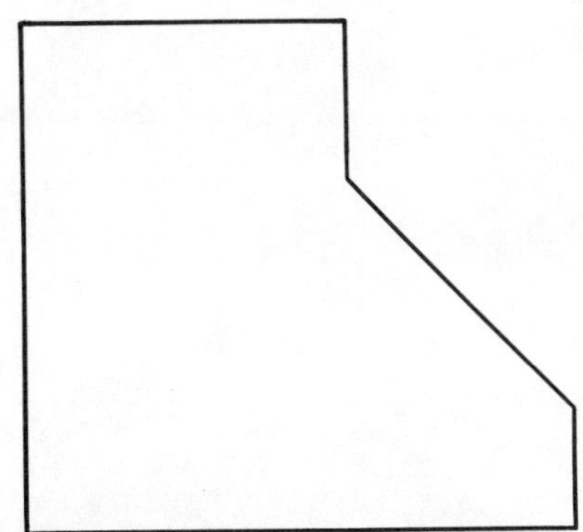

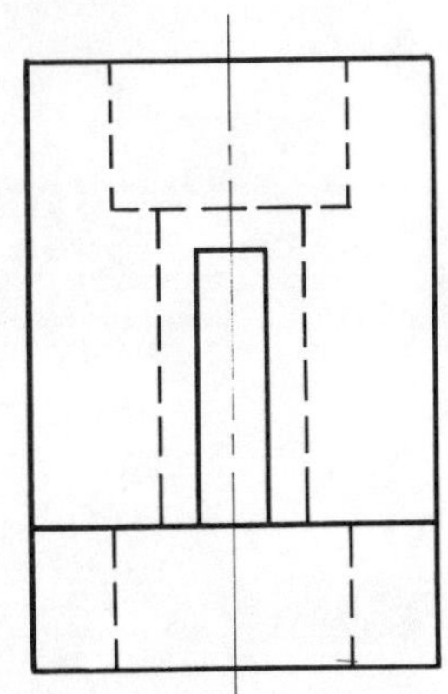

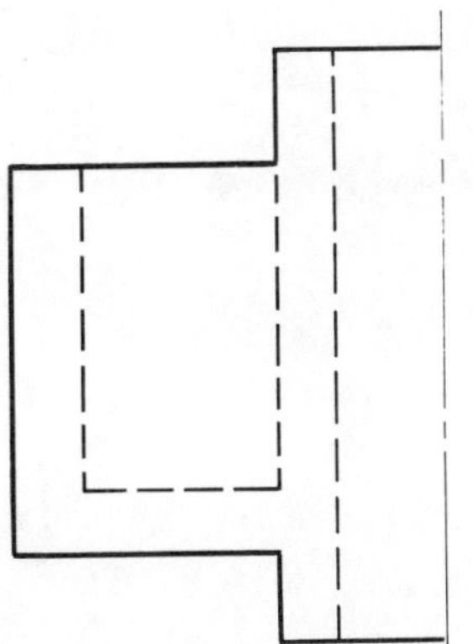

JHE

Graphics & Geometry ©	NAME FILE SEC DATE	MIN.	GRADE	44

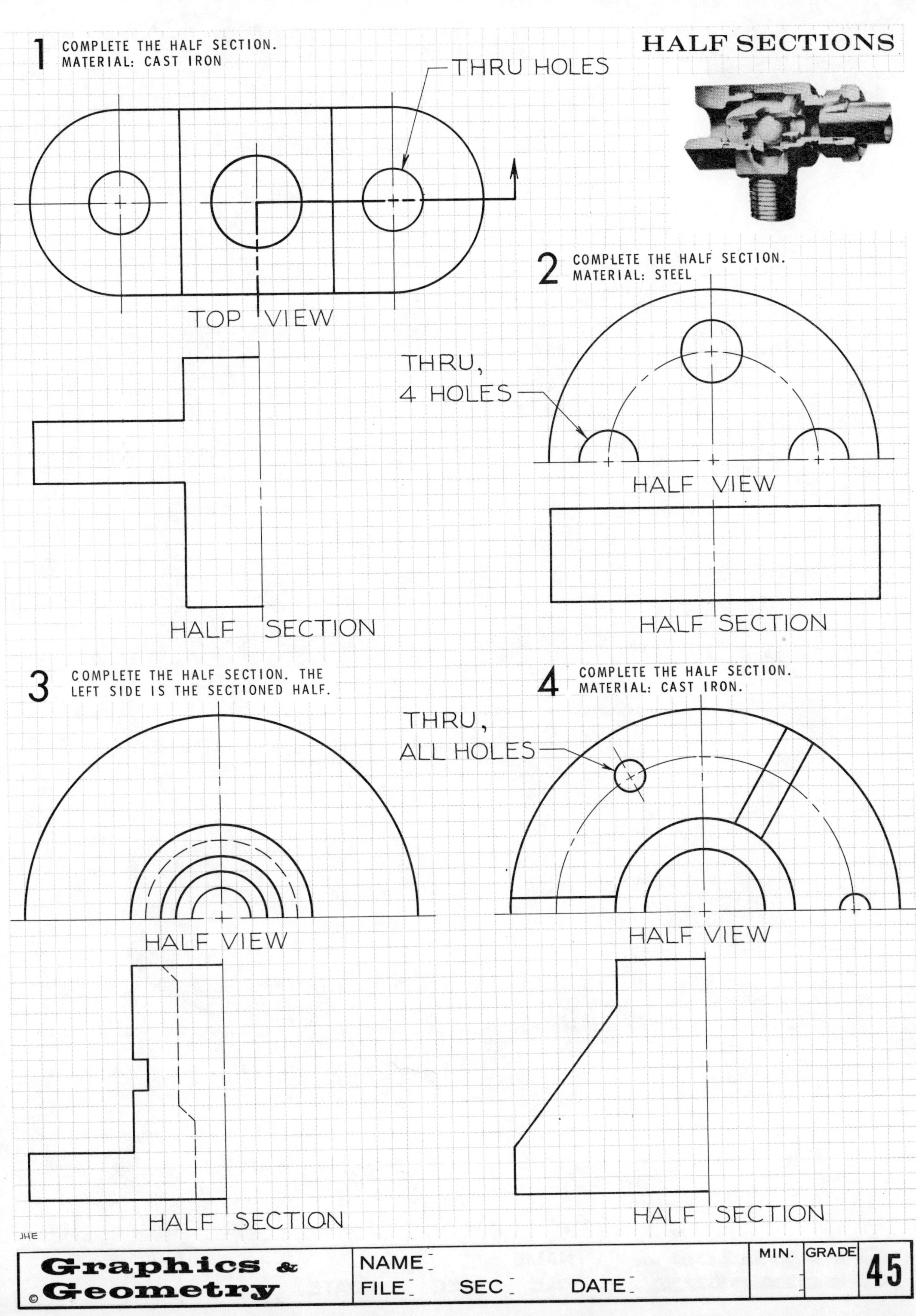

HALF SECTIONS
1 COMPLETE THE HALF SECTION.
MATERIAL: CAST IRON
THRU HOLES
TOP VIEW
HALF SECTION
2 COMPLETE THE HALF SECTION.
MATERIAL: STEEL
THRU, 4 HOLES
HALF VIEW
HALF SECTION
3 COMPLETE THE HALF SECTION. THE LEFT SIDE IS THE SECTIONED HALF.
HALF VIEW
HALF SECTION
4 COMPLETE THE HALF SECTION.
MATERIAL: CAST IRON.
THRU, ALL HOLES
HALF VIEW
HALF SECTION
JHE

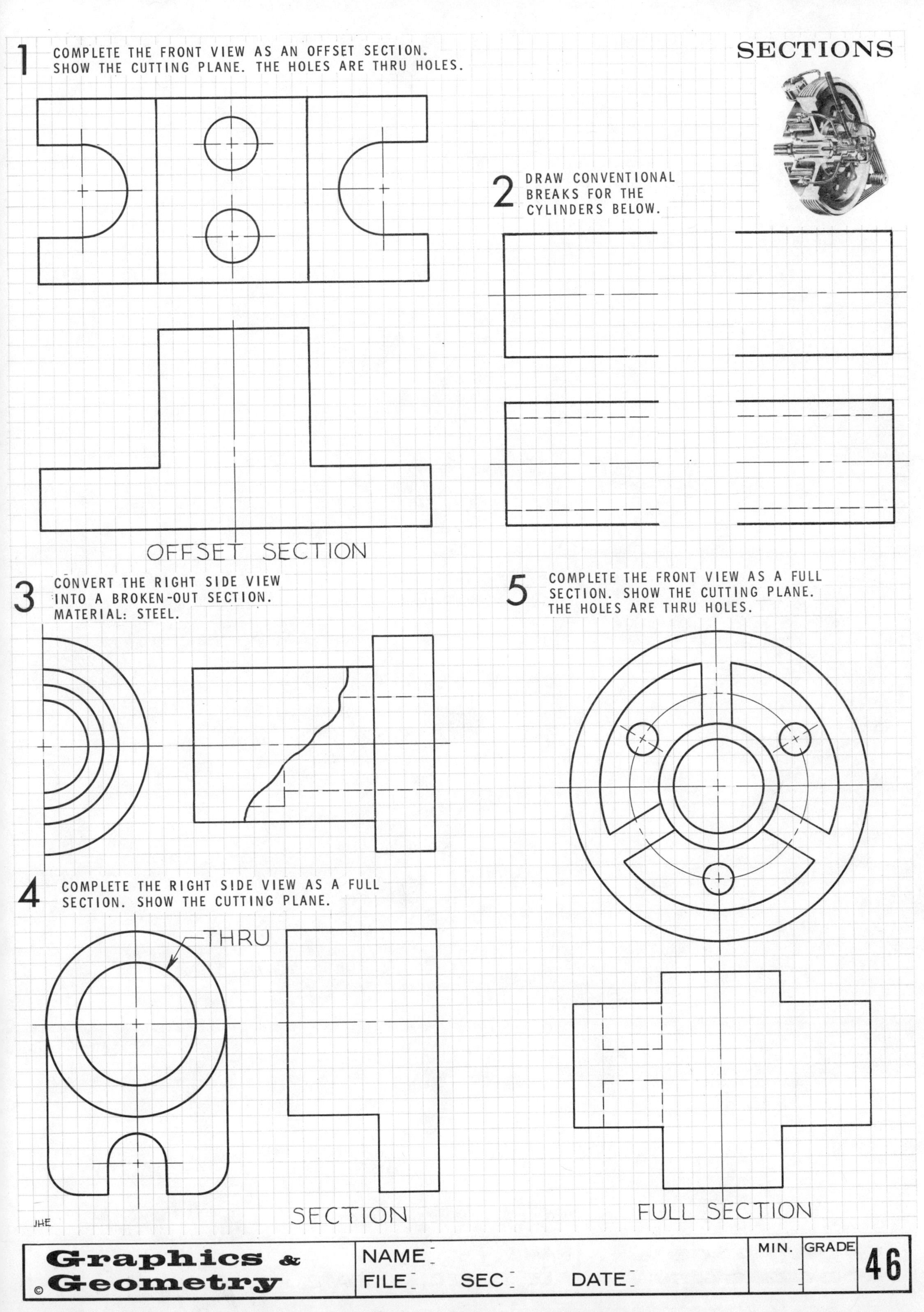
1
COMPLETE THE FRONT VIEW AS AN OFFSET SECTION.
SHOW THE CUTTING PLANE. THE HOLES ARE THRU HOLES.
SECTIONS
OFFSET SECTION
2
DRAW CONVENTIONAL
BREAKS FOR THE
CYLINDERS BELOW.
3
CONVERT THE RIGHT SIDE VIEW
INTO A BROKEN-OUT SECTION.
MATERIAL: STEEL.
5
COMPLETE THE FRONT VIEW AS A FULL
SECTION. SHOW THE CUTTING PLANE.
THE HOLES ARE THRU HOLES.
4
COMPLETE THE RIGHT SIDE VIEW AS A FULL
SECTION. SHOW THE CUTTING PLANE.
THRU
SECTION
FULL SECTION
JHE
Graphics & Geometry
©
NAME
FILE
SEC
DATE
MIN.
GRADE
46

Graphics & Geometry ©	NAME		MIN.	GRADE	
	FILE	SEC	DATE			

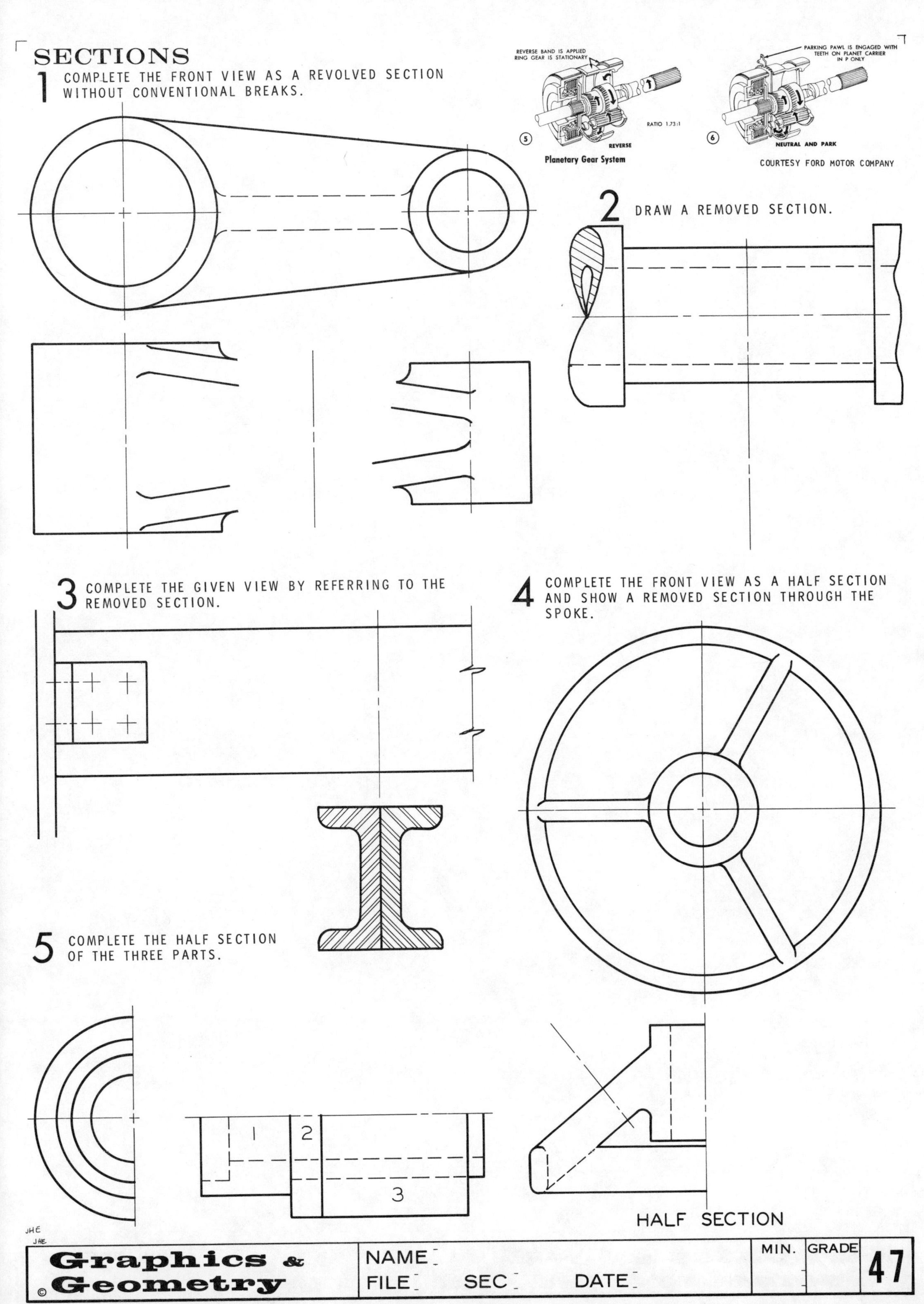
SECTIONS
1 COMPLETE THE FRONT VIEW AS A REVOLVED SECTION WITHOUT CONVENTIONAL BREAKS.
REVERSE BAND IS APPLIED RING GEAR IS STATIONARY
RATIO 1.73:1
5
REVERSE
Planetary Gear System
PARKING PAWL IS ENGAGED WITH TEETH ON PLANET CARRIER IN P ONLY
6
NEUTRAL AND PARK
COURTESY FORD MOTOR COMPANY
2 DRAW A REMOVED SECTION.
3 COMPLETE THE GIVEN VIEW BY REFERRING TO THE REMOVED SECTION.
4 COMPLETE THE FRONT VIEW AS A HALF SECTION AND SHOW A REMOVED SECTION THROUGH THE SPOKE.
5 COMPLETE THE HALF SECTION OF THE THREE PARTS.
1
2
3
HALF SECTION
JHE
Graphics & Geometry
©
NAME
FILE
SEC
DATE
MIN.
GRADE
47

Graphics & © Geometry	NAME:	MIN.	GRADE	
	FILE: SEC: DATE:			

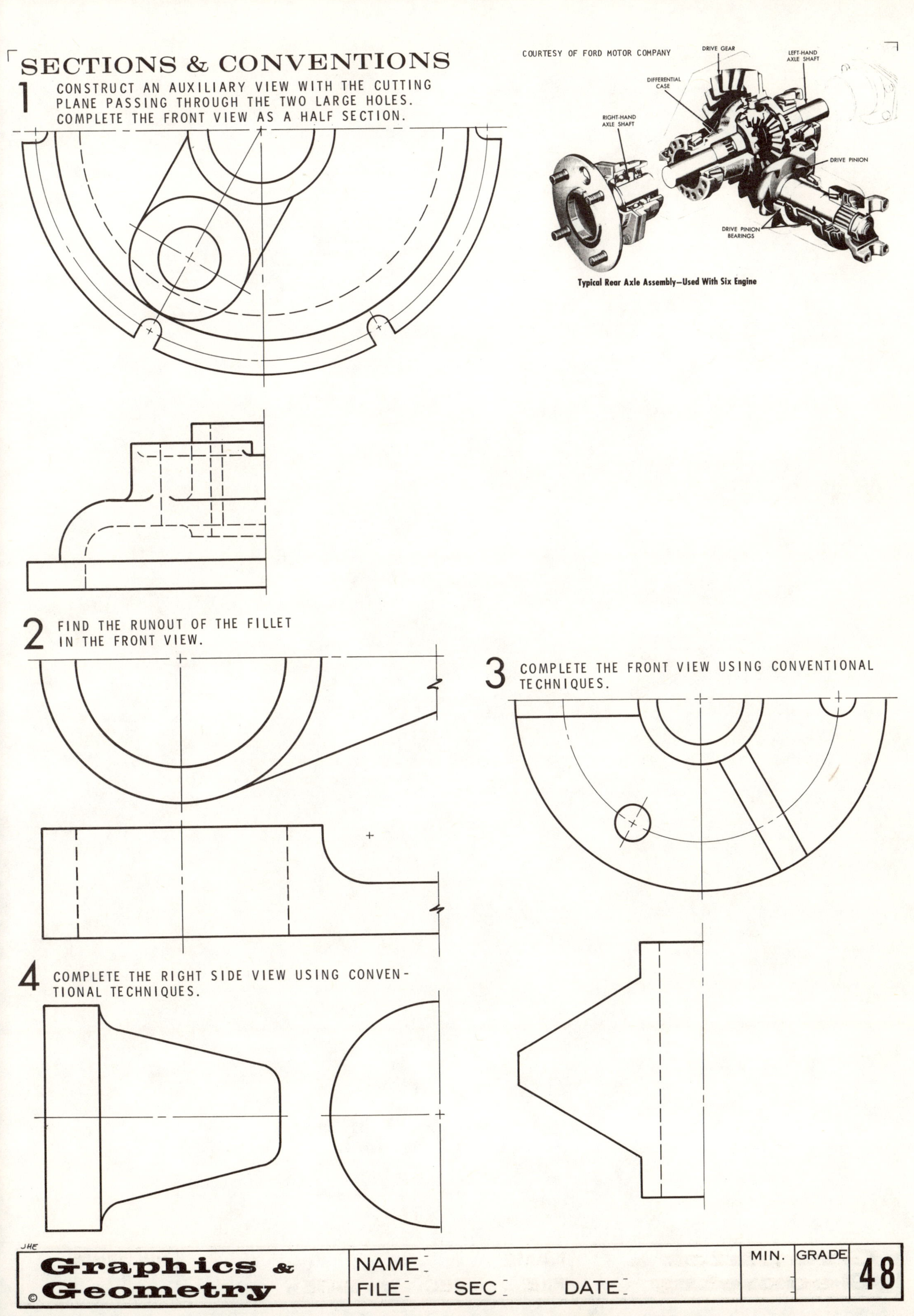
SECTIONS & CONVENTIONS
1 CONSTRUCT AN AUXILIARY VIEW WITH THE CUTTING PLANE PASSING THROUGH THE TWO LARGE HOLES. COMPLETE THE FRONT VIEW AS A HALF SECTION.
COURTESY OF FORD MOTOR COMPANY
DRIVE GEAR
LEFT-HAND AXLE SHAFT
DIFFERENTIAL CASE
RIGHT-HAND AXLE SHAFT
DRIVE PINION
DRIVE PINION BEARINGS
Typical Rear Axle Assembly—Used With Six Engine
2 FIND THE RUNOUT OF THE FILLET IN THE FRONT VIEW.
3 COMPLETE THE FRONT VIEW USING CONVENTIONAL TECHNIQUES.
4 COMPLETE THE RIGHT SIDE VIEW USING CONVENTIONAL TECHNIQUES.
JHE
Graphics & Geometry
©
NAME
FILE
SEC
DATE
MIN.
GRADE
48

Graphics & Geometry	NAME			MIN.	GRADE	
	FILE	SEC	DATE			

DRAW THE MISSING VIEWS OF EACH LINE AND INDICATE WHAT TYPE OF LINE EACH IS. LABEL TRUE LENGTH VIEWS TL.

1 TYPE:

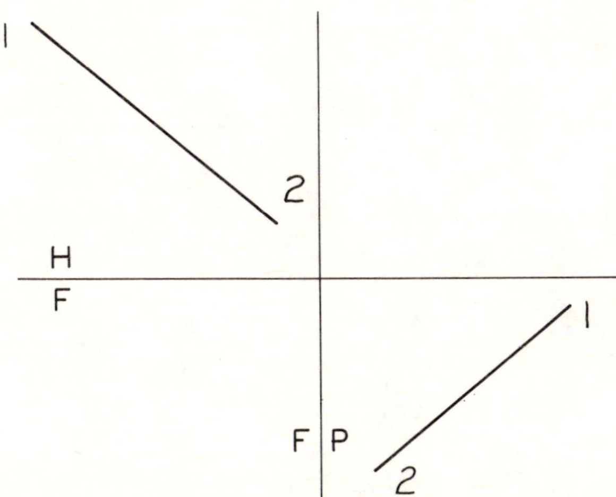

2 TYPE:

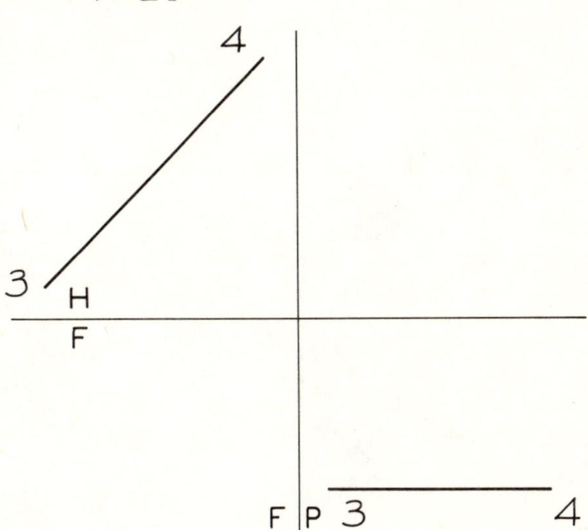

3 TYPE:

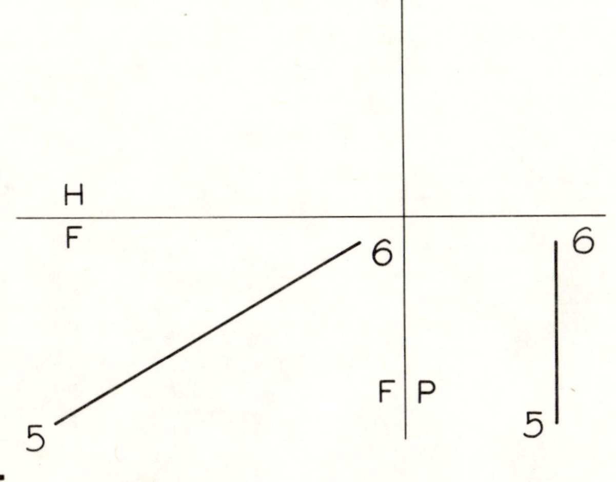

4 TYPE:

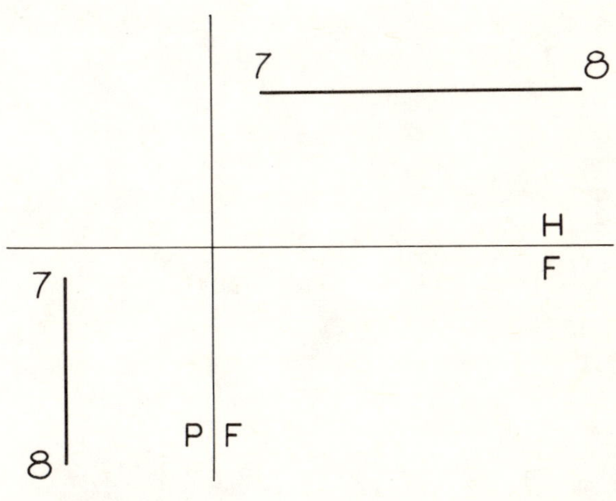

5 TYPE:

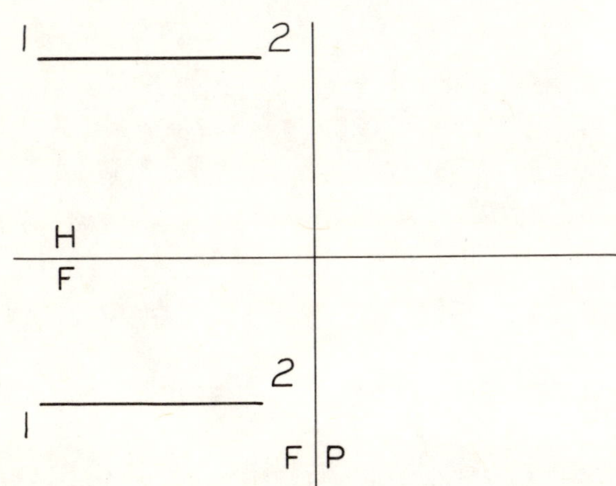

6 TYPE:

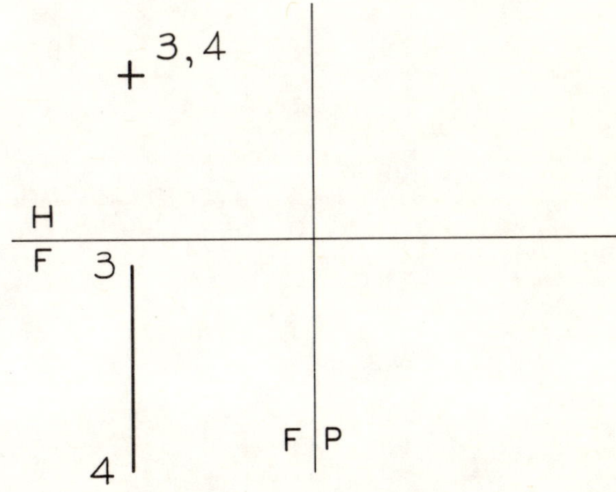

G R

Graphics & Geometry	NAME:	MIN.	GRADE	49
©	FILE: SEC: DATE:			

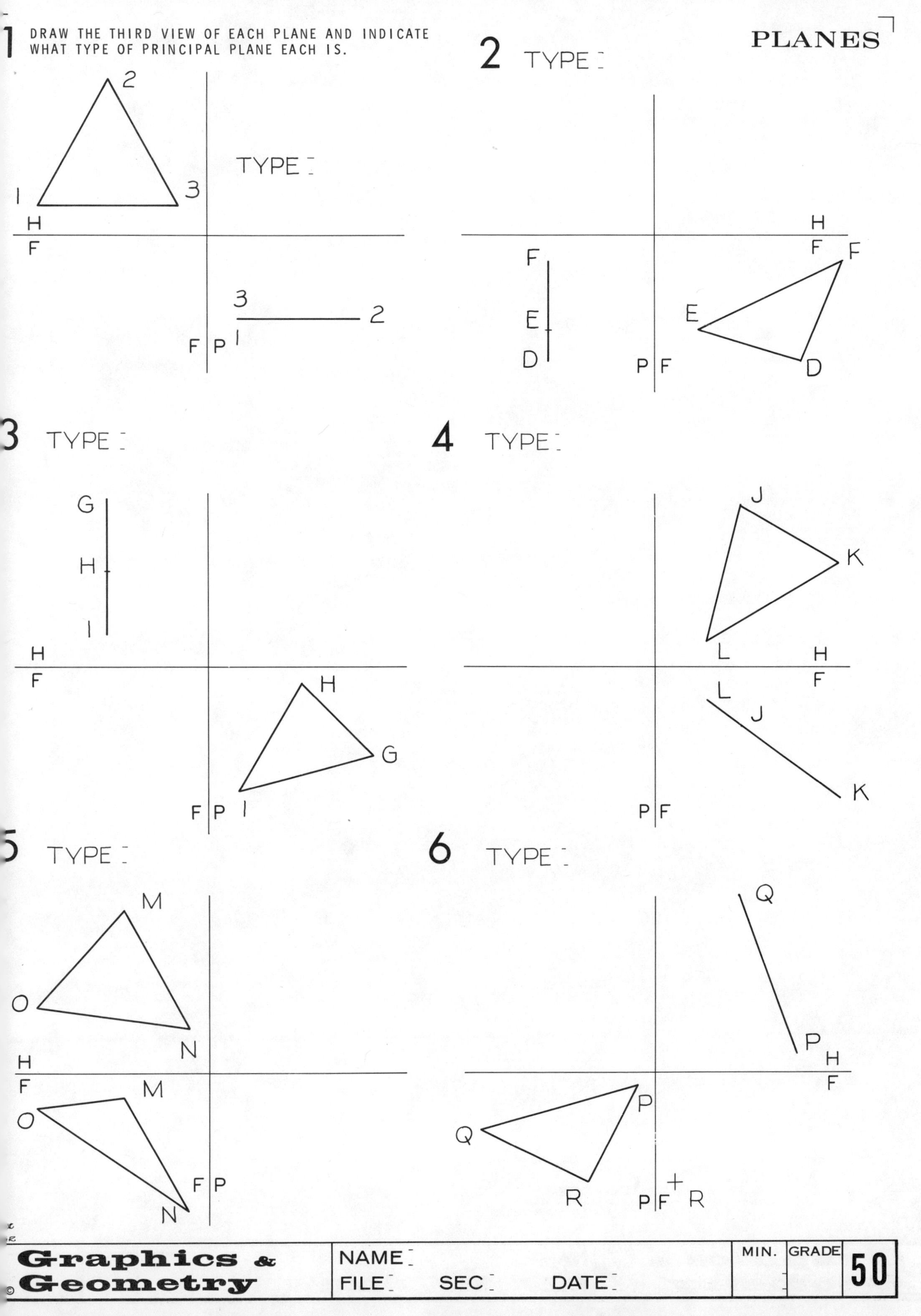
PLANES
1 DRAW THE THIRD VIEW OF EACH PLANE AND INDICATE WHAT TYPE OF PRINCIPAL PLANE EACH IS.
TYPE
2 TYPE
3 TYPE
4 TYPE
5 TYPE
6 TYPE
H
F
P
1
2
3
D
E
F
G
H
I
J
K
L
M
N
O
P
Q
R
Graphics & Geometry
NAME
FILE
SEC
DATE
MIN.
GRADE
50

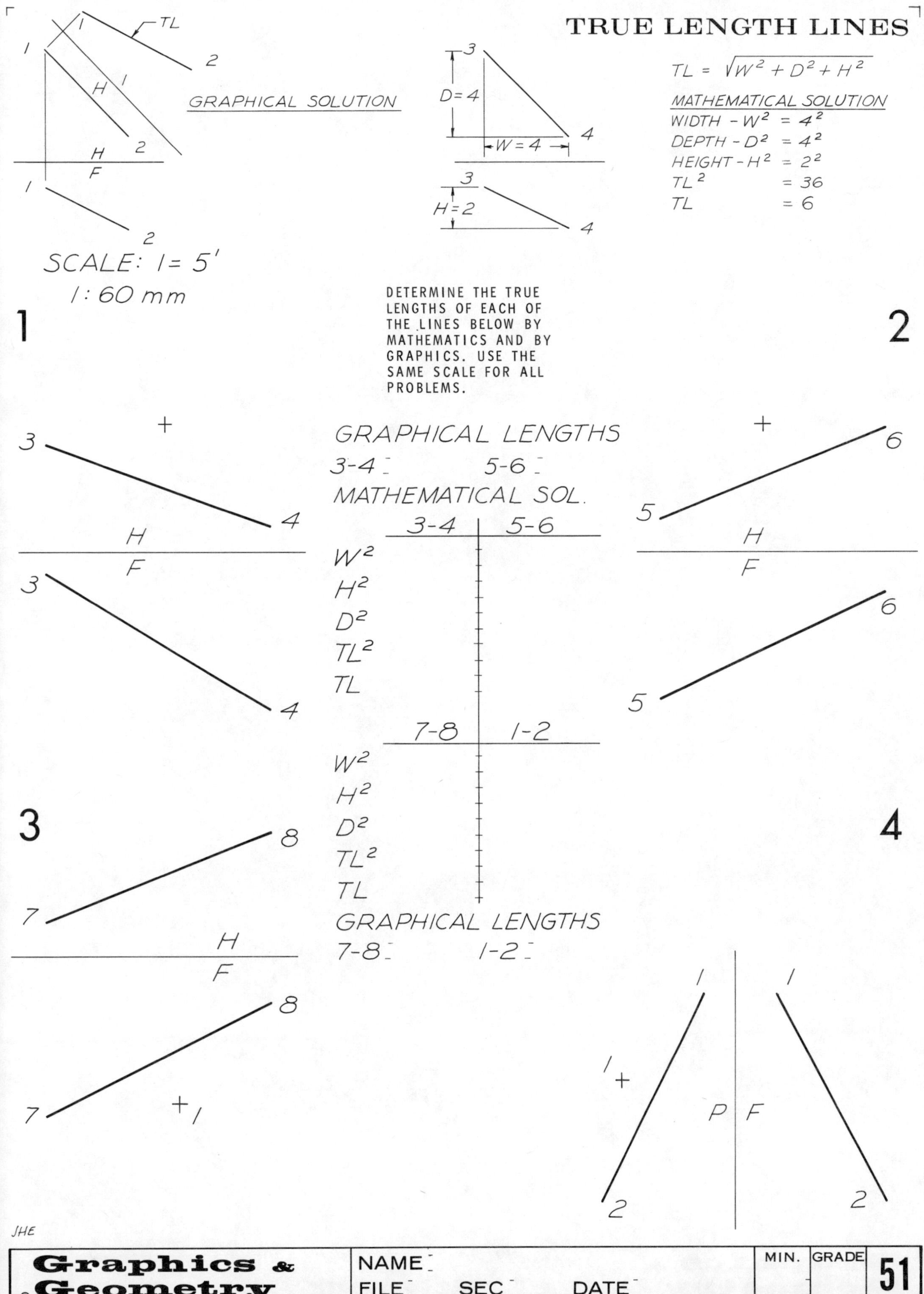
TRUE LENGTH LINES
TL
GRAPHICAL SOLUTION
H
F
D=4
W=4
H=2
$TL = \sqrt{W^2 + D^2 + H^2}$
MATHEMATICAL SOLUTION
WIDTH - W^2 = 4^2
DEPTH - D^2 = 4^2
HEIGHT - H^2 = 2^2
TL^2 = 36
TL = 6
SCALE: 1= 5'
1: 60 mm
1
2
DETERMINE THE TRUE LENGTHS OF EACH OF THE LINES BELOW BY MATHEMATICS AND BY GRAPHICS. USE THE SAME SCALE FOR ALL PROBLEMS.
GRAPHICAL LENGTHS
3-4
5-6
MATHEMATICAL SOL.
3-4
5-6
W^2
H^2
D^2
TL^2
TL
7-8
1-2
W^2
H^2
D^2
TL^2
TL
GRAPHICAL LENGTHS
7-8
1-2
3
4
P
F
JHE
Graphics & Geometry
NAME
FILE
SEC
DATE
MIN.
GRADE
51

Graphics & Geometry	NAME	MIN.	GRADE	
©	FILE SEC DATE			

TRUE LENGTH LINES

BEFORE A DETAILED SET OF PLANS CAN BE PREPARED FOR CONSTRUCTING A HELICOPTER FRAME, ITS GEOMETRY MUST BE DETERMINED. (THE GIVEN VIEWS OF THE FRAME HAVE BEEN SIMPLIFIED TO CLARIFY THE PROBLEMS BELOW.)

1 FIND EACH OF THE LINES MENTIONED IN THE TABLE TRUE LENGTH. MEASURE FROM THE INTERSECTIONS OF THEIR CENTERLINES. PROJECT LINE 1-2 FROM THE TOP VIEW AND LINE 2-3 FROM THE FRONT VIEW.

2 FIND THE TRUE LENGTH VIEW OF EACH LINE MENTIONED IN THE TABLE BY THE TRUE-LENGTH DIAGRAM METHOD. LAY OFF THE DIMENSIONS AS SPECIFIED BY THE TRUE LENGTH DIAGRAM.

VERTICAL DISTANCES

SCALE: 1 = 2'
1:25 mm

HORIZ. DISTANCES
TL DIAGRAM

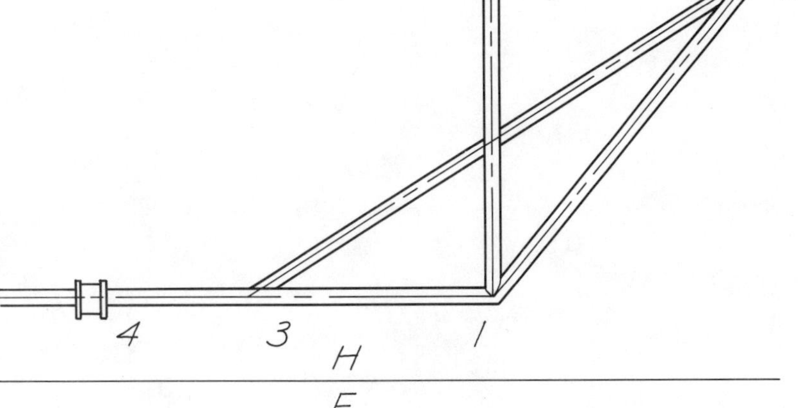

H
F

1
4
2
3

+ 1

+ 1

MEMBER	TRUE L.
1-2	
2-3	
3-4	
1-3	

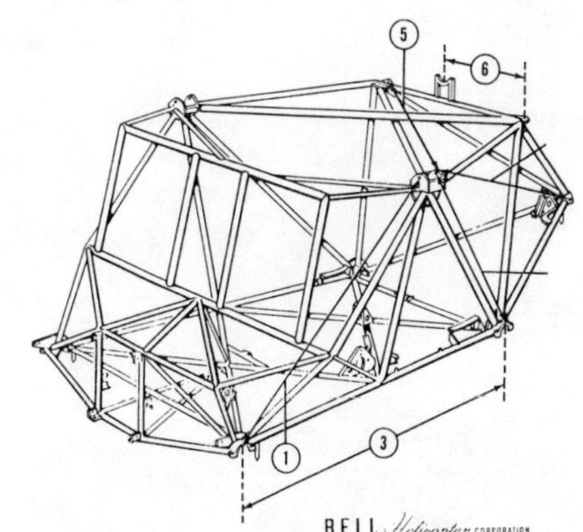

JHE

Graphics & Geometry ©	NAME FILE SEC DATE	MIN.	GRADE	52

METRIC SCALES

Remove this page and fold it longwise along the desired scale for making metric measurements.

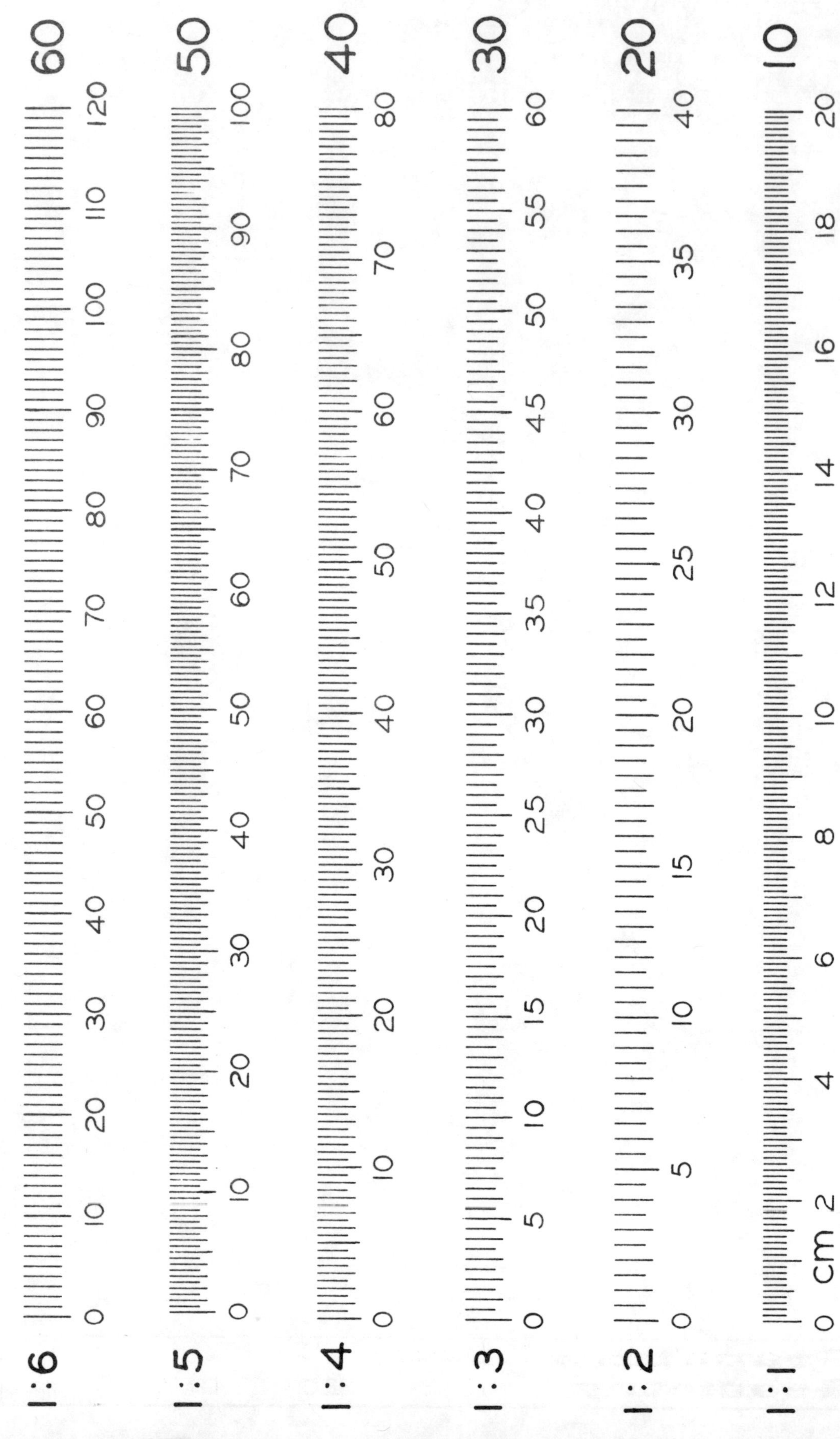

SCALE: 1 = 5'
1:60

1 ANALYZING THE GIVEN LINE FROM 1 TO 2, COMPLETE THE TABLE OF VALUES BELOW.

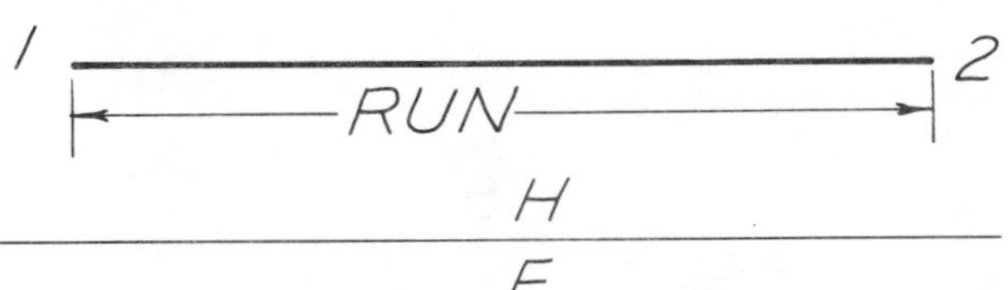

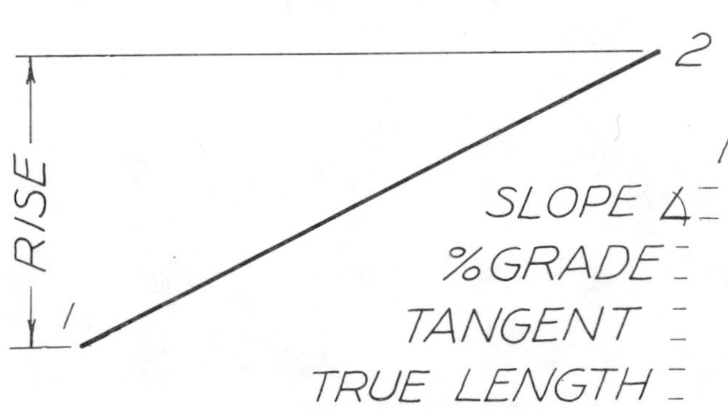

SLOPE ∠ ___
% GRADE ___
TANGENT ___
TRUE LENGTH ___

2 DRAW A FRONTAL LINE 12.5' OR 762 mm LONG DOWNWARD AT 30° FROM POINT 3 TO POINT 4. SHOW THE TOP AND FRONT VIEWS AND COMPLETE THE TABLE.

3 +

H
F

3+

1
2

3 USING THE AUXILIARY VIEW METHOD, COMPLETE THE TABLE OF VALUES BELOW.

4 DRAW LINE 8-7 WITH A -30% GRADE FROM POINT 8. COMPLETE THE FRONT VIEW AND THE TABLE OF VALUES. SOLVE BY AUXILIARY VIEW METHOD.

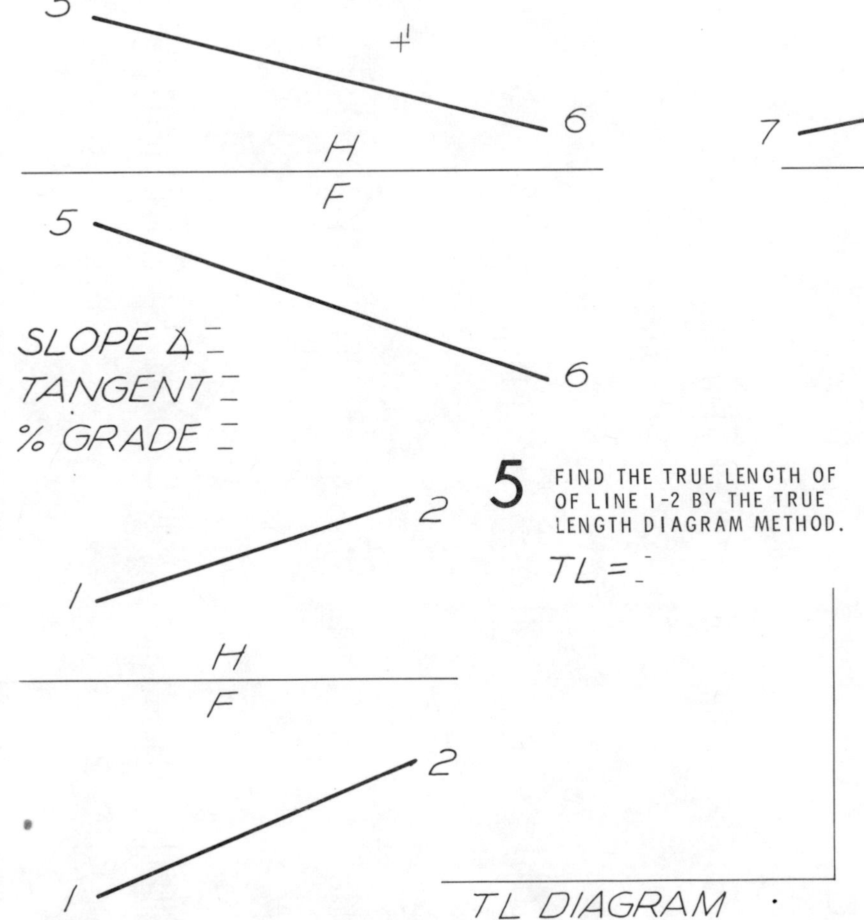

SLOPE ∠ ___
TANGENT ___
% GRADE ___

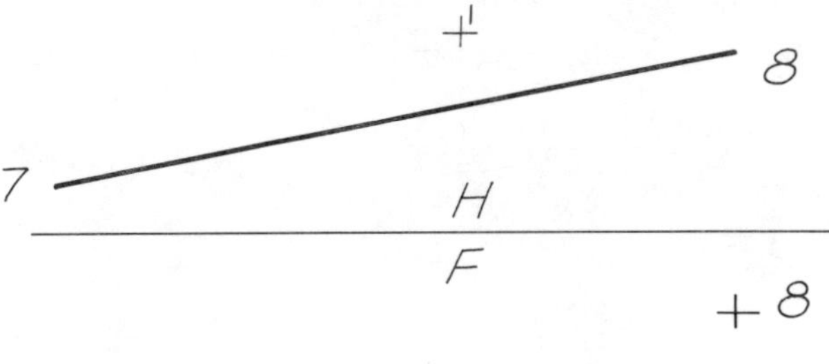

SLOPE ∠ ___
TANGENT ___
LENGTH ___

5 FIND THE TRUE LENGTH OF OF LINE 1-2 BY THE TRUE LENGTH DIAGRAM METHOD.

TL = ___

TL DIAGRAM

COURTESY OF AMERICAN AGGREGATES CORPORATION

JHE-LES

Graphics & Geometry ©	NAME ___ FILE ___ SEC ___ DATE ___	MIN.	GRADE	53

METRIC SCALES

Remove this page and fold it longwise along the desired scale for making metric measurements.

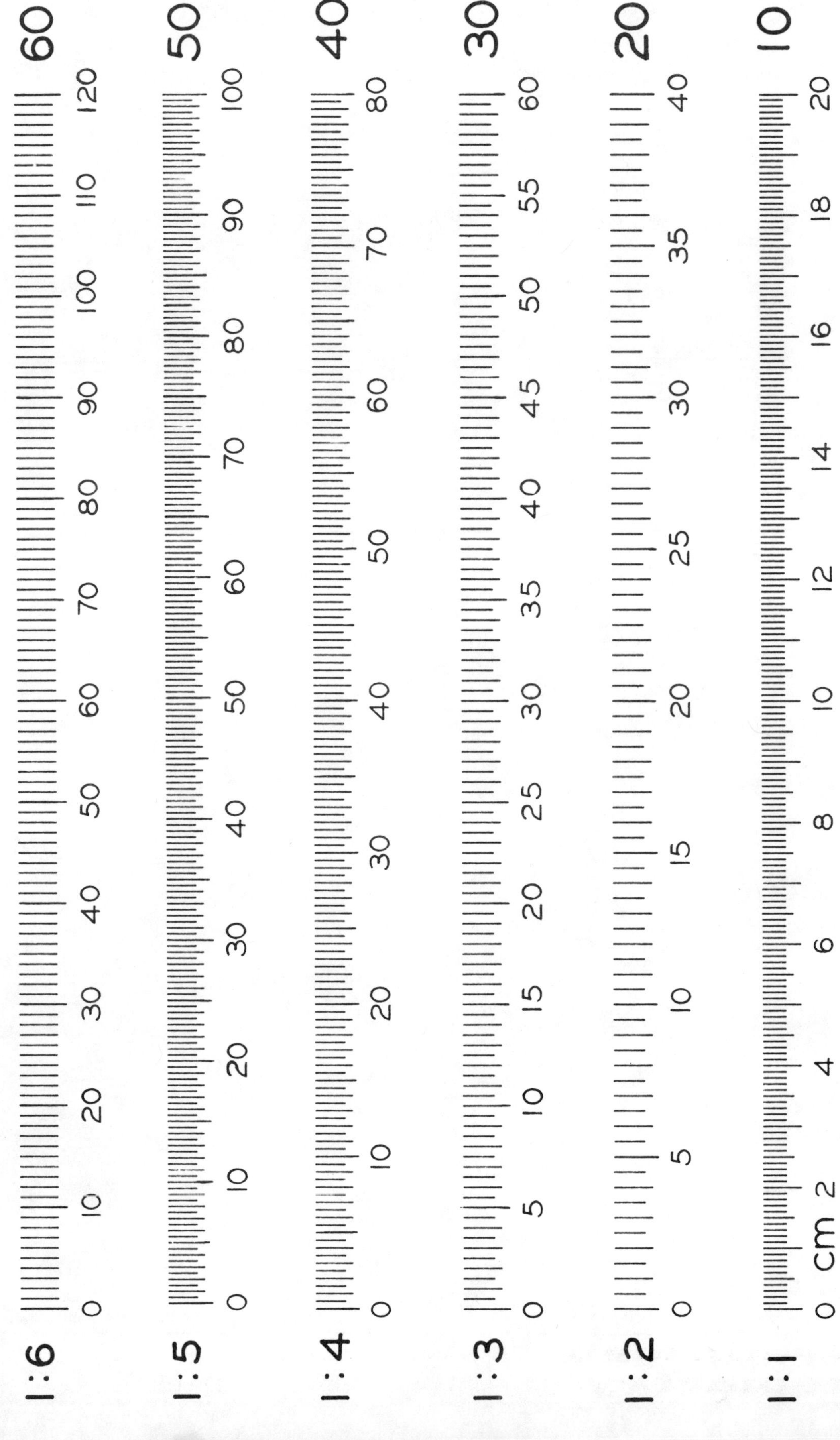

SLOPE AND BEARING

1 DETERMINE THE COMPASS AND AZIMUTH BEARINGS OF THE BOUNDARY LINES OF THE PLOT. BEGIN AT POINT 1.

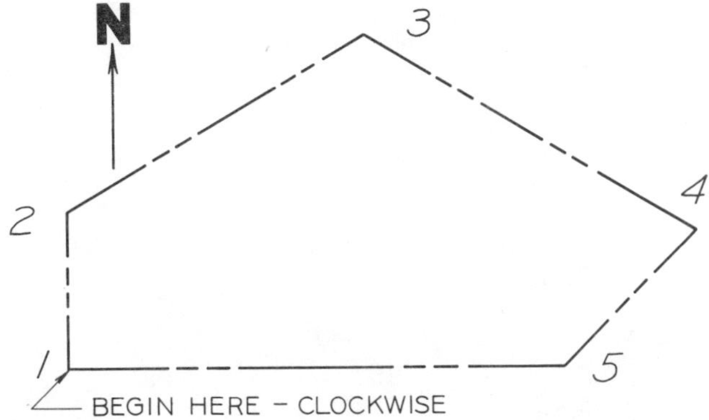

LINE	BEARING	AZIMUTH
1-2		
2-3		
4-5		
5-1		

2 DETERMINE THE BEARING, PER CENT GRADE AND SLOPE OF THE LINE FROM 6 TO 7 BY THE AUXILIARY METHOD.

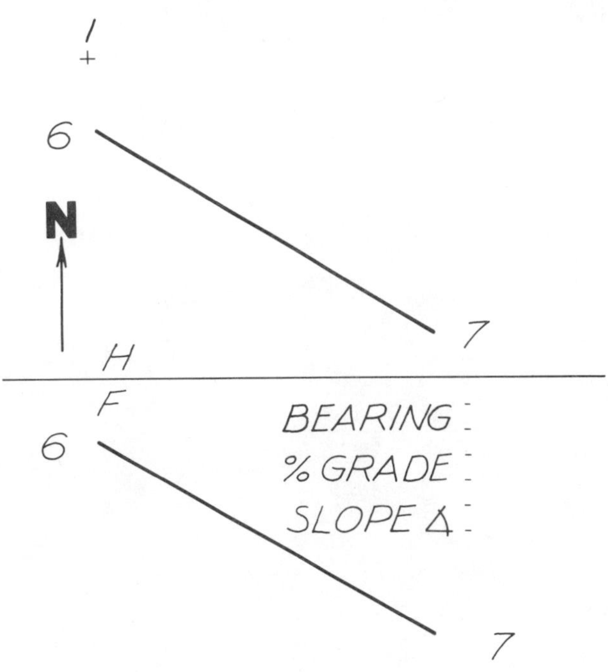

BEARING

% GRADE

SLOPE ∠

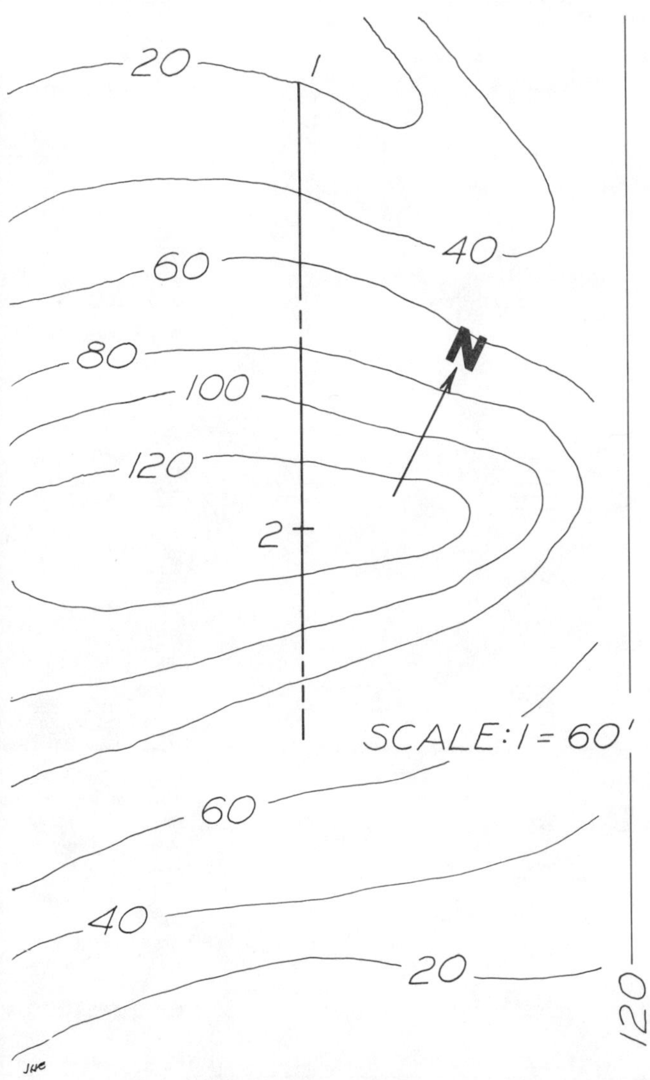

3 A TUNNEL IS TO BE DRILLED FROM POINT 1 THROUGH 2 TO 3. LINE 1-2 REPRESENTS THE BOTTOM OF THE TUNNEL THAT HAS A +20% GRADE. THE TUNNEL IS LEVEL FROM POINT 2 TO POINT 3 WHICH LIES ON THE SURFACE. FIND A VERTICAL SECTION THROUGH 1-2-3 AND SHOW THE TUNNEL. LOCATE POINT 3 IN THE PLAN VIEW.

LINE	BEARING	LENGTH
1-2		
2-3		

Courtesy of U. S. Dept of Interior

JHE

METRIC SCALES

Remove this page and fold it longwise along the desired scale for making metric measurements.

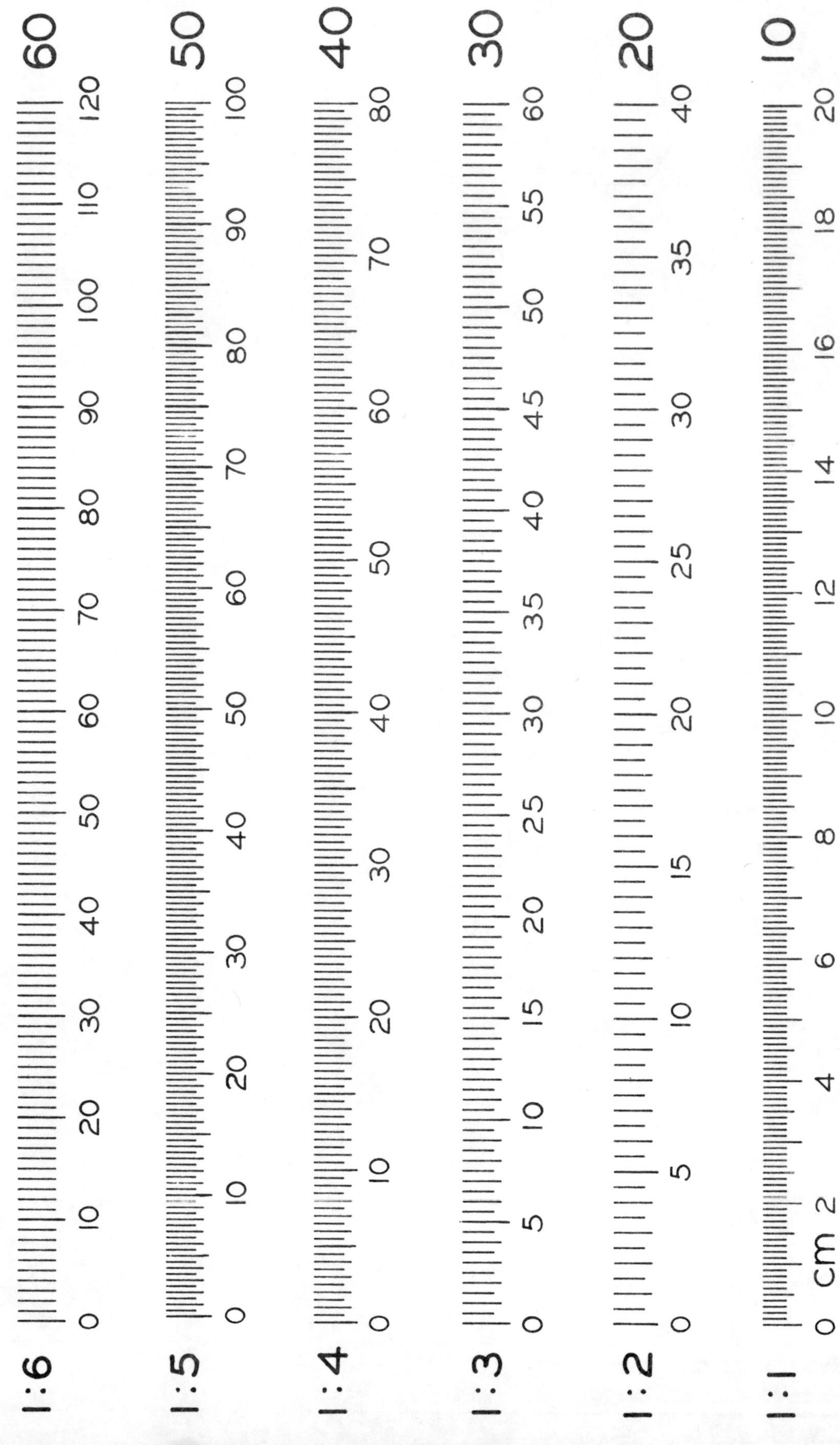

EDGE VIEW OF A PLANE

1 FIND THE EDGE VIEW OF THE PLANE AS INDICATED BY THE SIGHT ARROW.

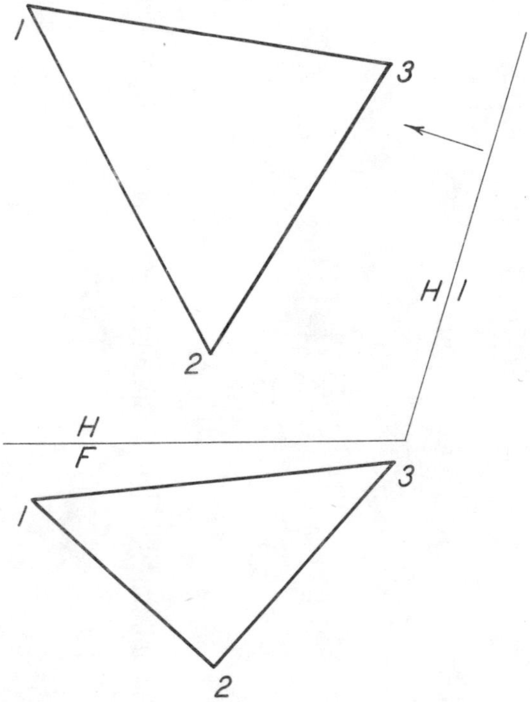
1
3
2
H|1
H
F
1
3
2

2 DRAW THE EDGE VIEW OF PLANE 1-2-3. CUT ALONG THE DASHED LINES AND FOLD ALONG THE REFERENCE PLANE LINES TO FORM A SPATIAL MODEL.

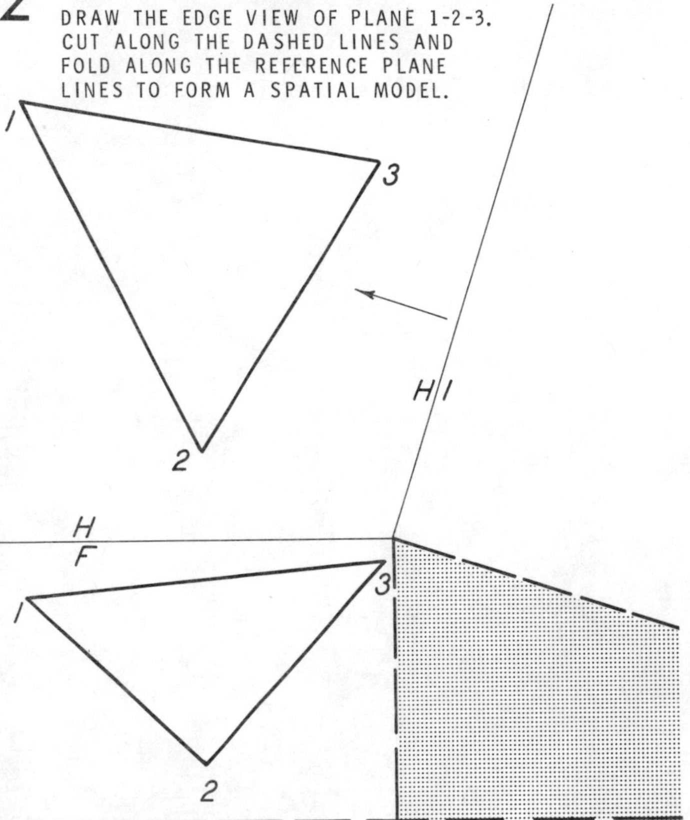
1
3
2
H|1
H
F
1
3
2

3 FIND THE EDGE VIEW OF THE PLANE AS INDICATED BY THE SIGHT ARROW.

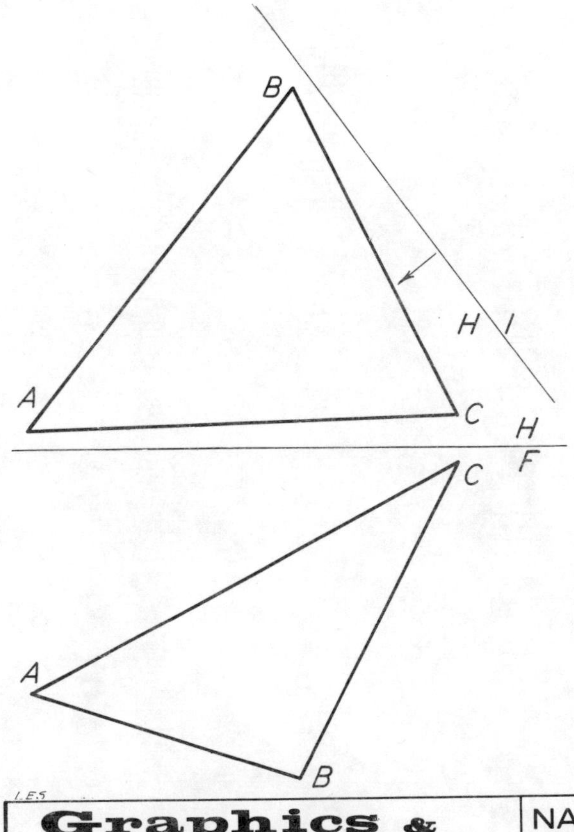
B
H|1
A
C
H
F
C
A
B

4 FIND THE EDGE VIEW OF THE PLANE AS INDICATED BY THE SIGHT ARROW. CUT ALONG THE DASHED LINES AND FOLD ALONG THE REFERENCE PLANE LINES HF AND F-1.

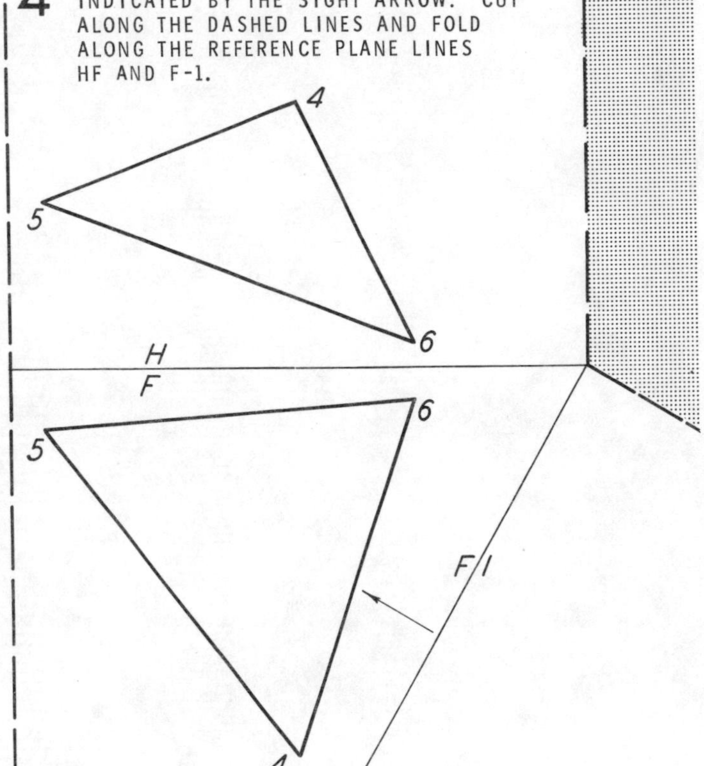
4
5
6
H
F
6
5
F|1
4

I.E.S

Graphics & Geometry ©	NAME FILE SEC DATE	MIN.	GRADE	55

METRIC SCALES

Remove this page and fold it longwise along the desired scale for making metric measurements.

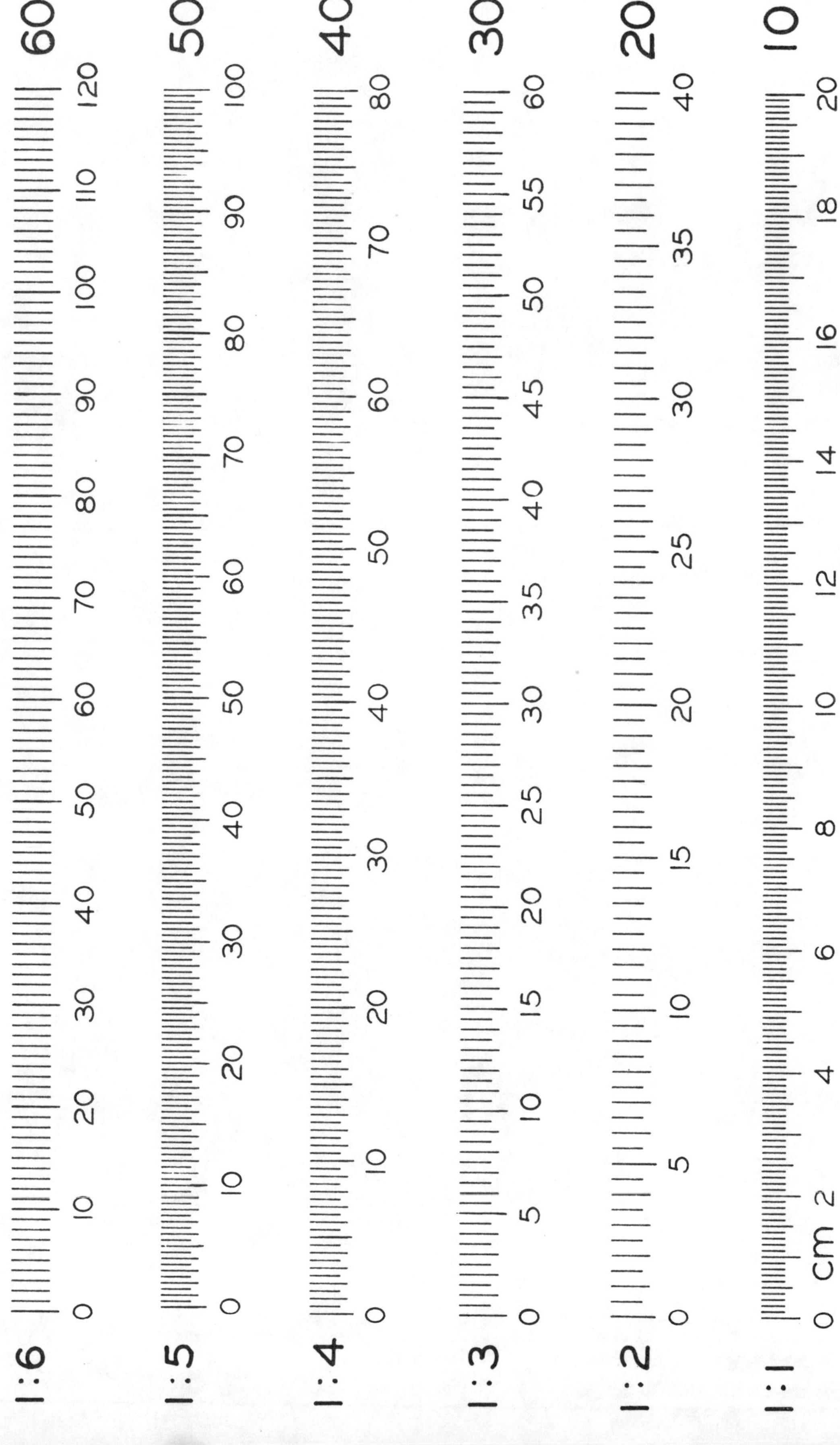

PIERCING POINTS

1 FIND THE PIERCING POINT OF LINE AB ON THE PLANE.

1 2 B A H 4 3 F 4,1 A B 3,2

2 FIND THE PIERCING POINT BY THE AUXILIARY VIEW METHOD. SHOW VISIBILITY IN ALL VIEWS.

A 6 +1 5 7 B 8 H F B 6 5 A 8 7

3 FIND THE PERPENDICULAR FROM O TO THE PLANE BY AN AUXILIARY VIEW. SHOW VISIBILITY IN ALL VIEWS.

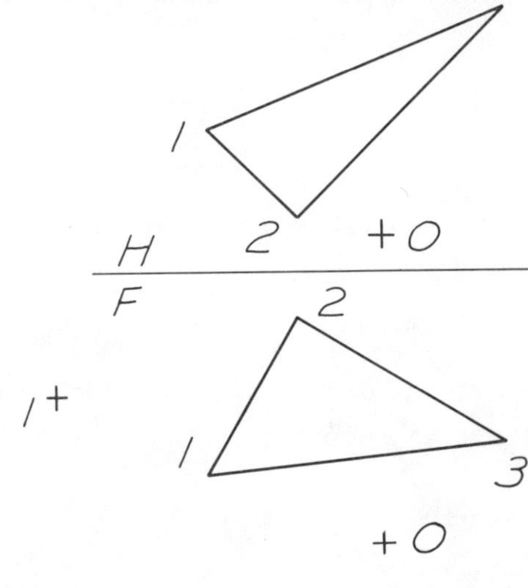

4 A CONTROL CABLE MUST PASS THROUGH A SUPPORT BRACKET OF AN AIRCRAFT. FIND THE PIERCING POINT AND SHOW THE VISIBILITY IN ALL VIEWS BY THE AUXILIARY VIEW METHOD BY PROJECTING FROM THE FRONT VIEW. CHECK YOUR SOLUTION BY THE CUTTING PLANE METHOD.

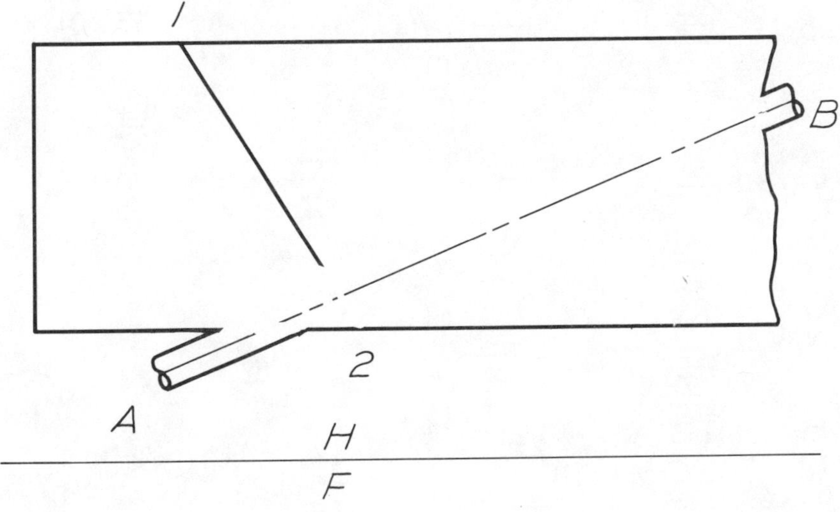

+1

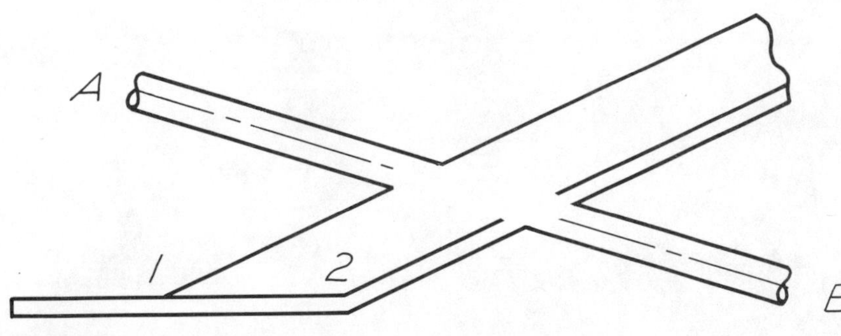

JHE

METRIC SCALES

Remove this page and fold it longwise along the desired scale for making metric measurements.

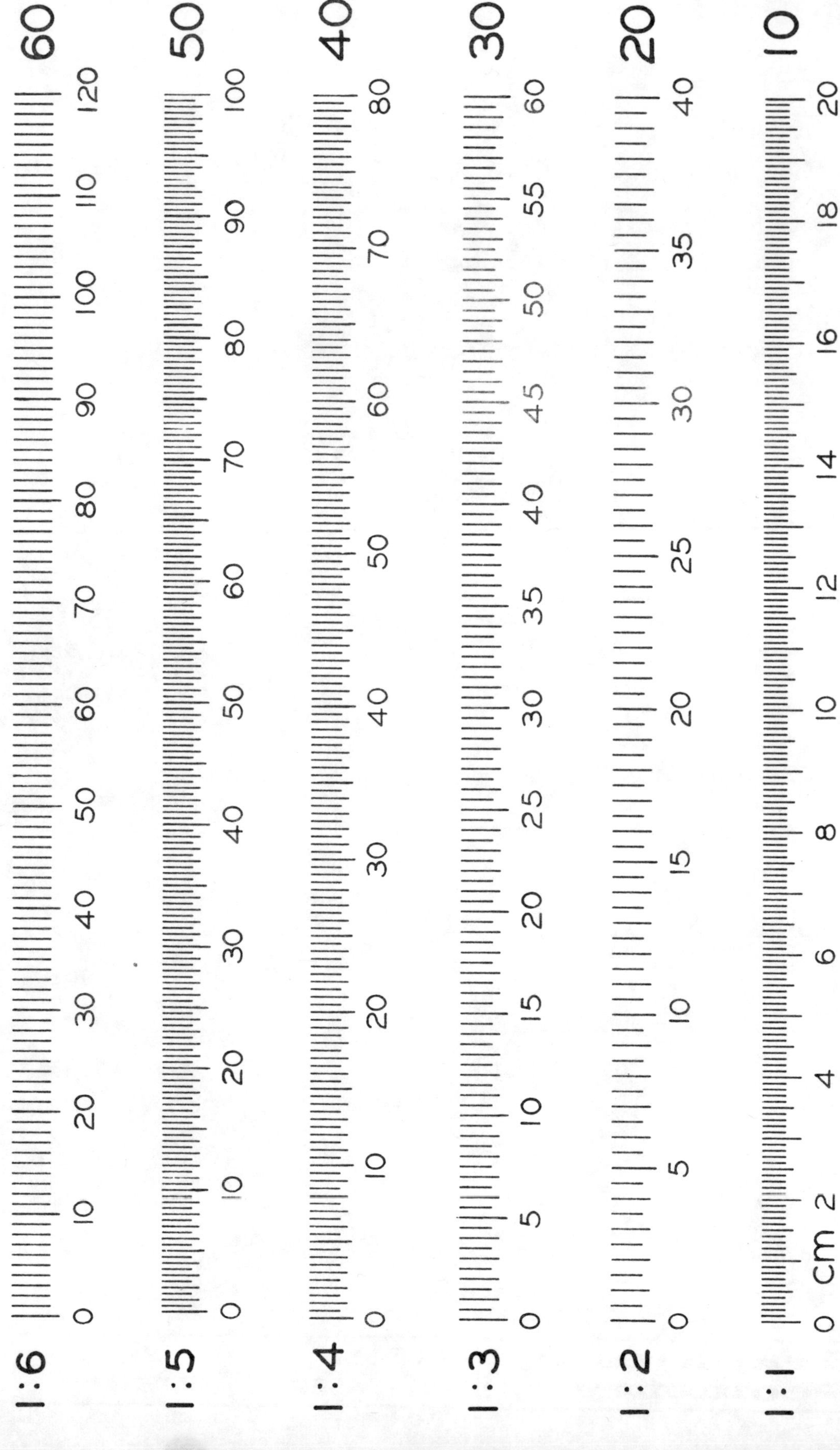

FIND THE SLOPE AND THE COMPASS DIRECTION OF THE SLOPE OF EACH PLANE IN PROBLEMS 1, 2, & 3. USE THE SAME NORTH ARROW FOR EACH PROBLEM.

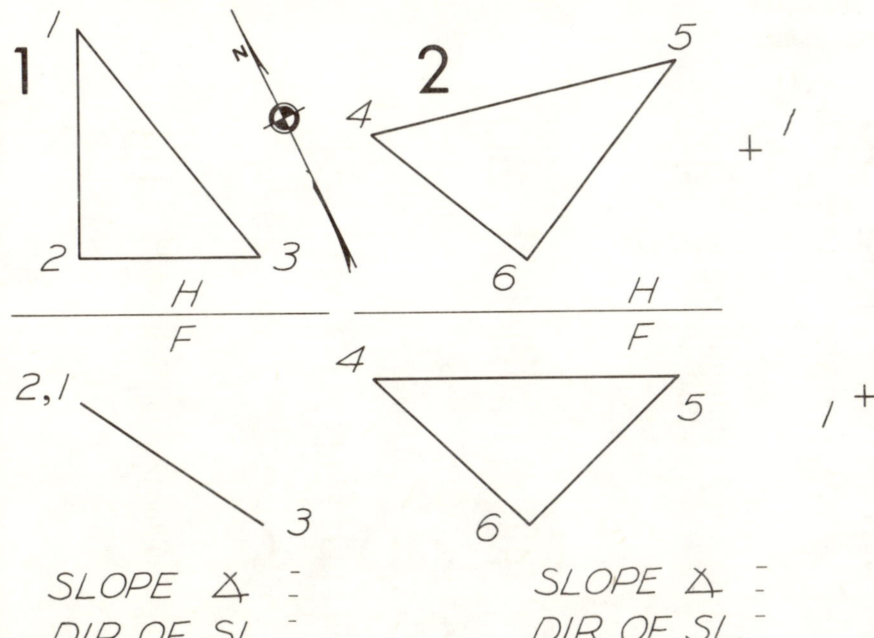

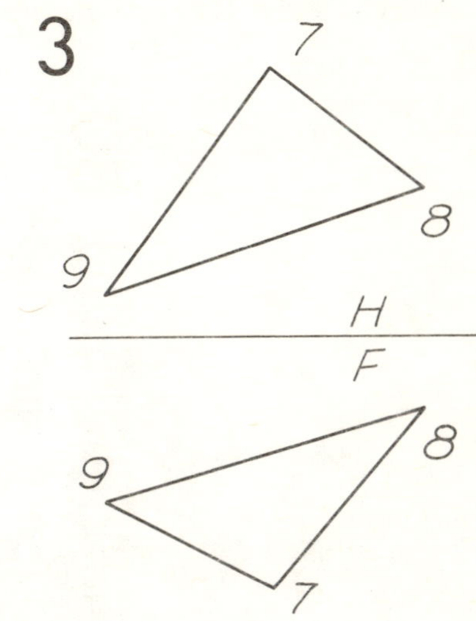

4 COMPLETE THE FRONT VIEW OF PLANE 1-2-3 USING THE FOLLOWING SPECIFICATIONS:

COMPASS DIRECTION OF SLOPE: S 20° E
SLOPE ANGLE: 40°

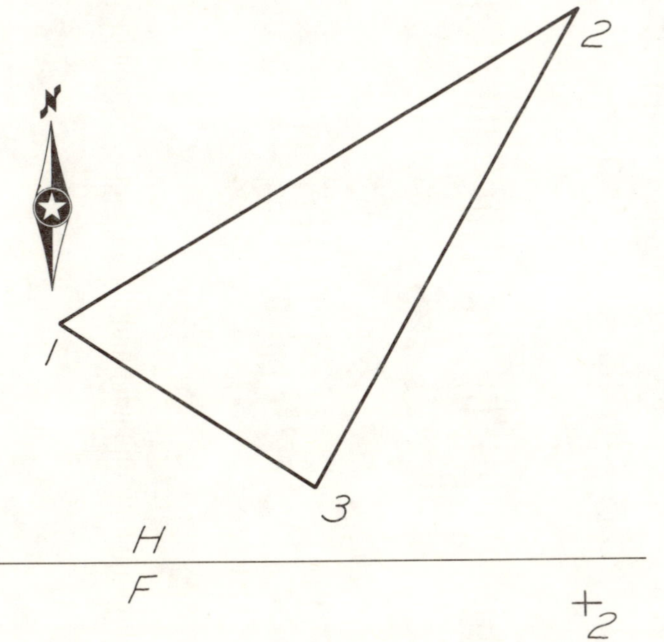

Courtesy U. S. Dept of Interior

JHE

METRIC SCALES

Remove this page and fold it longwise along the desired scale for making metric measurements.

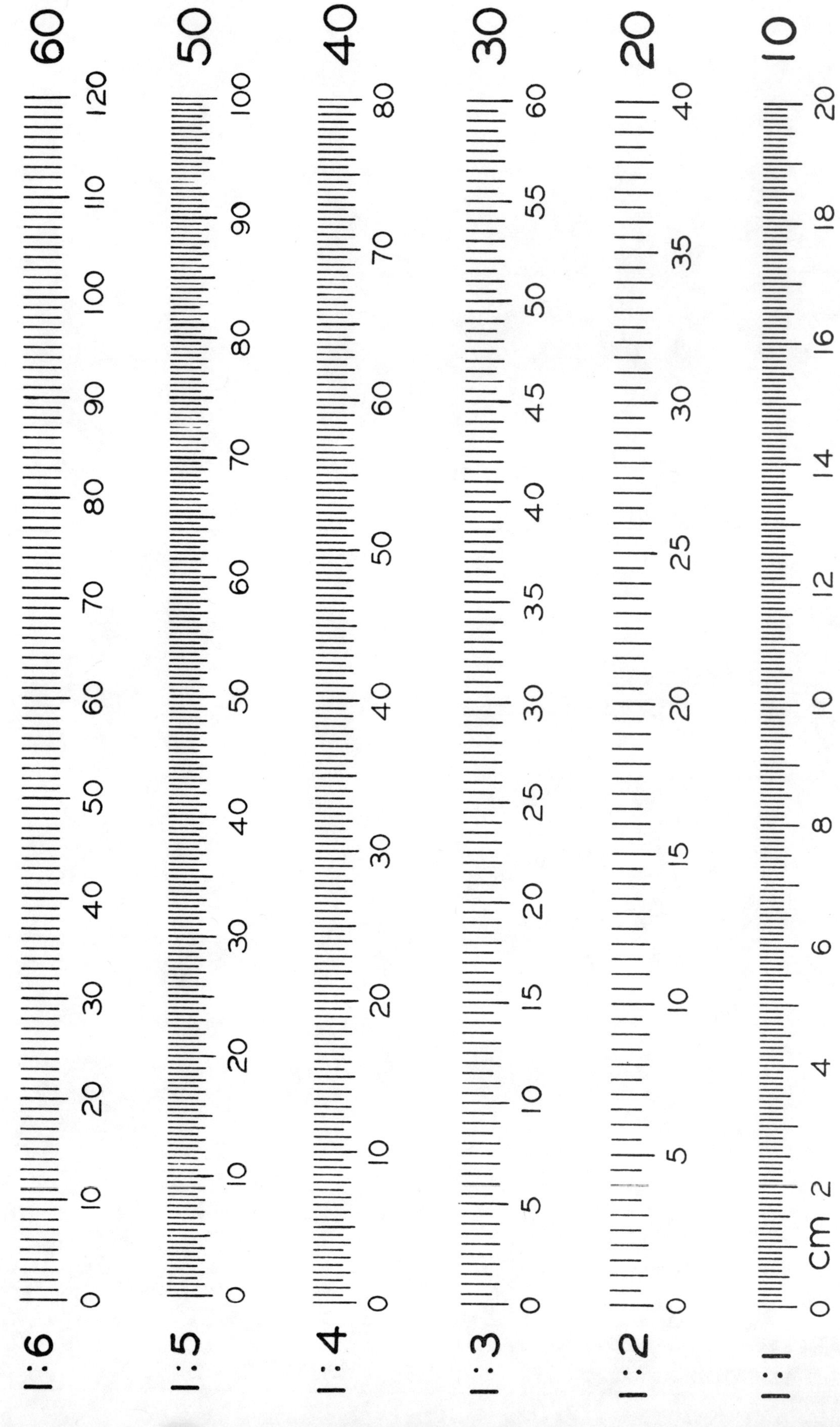

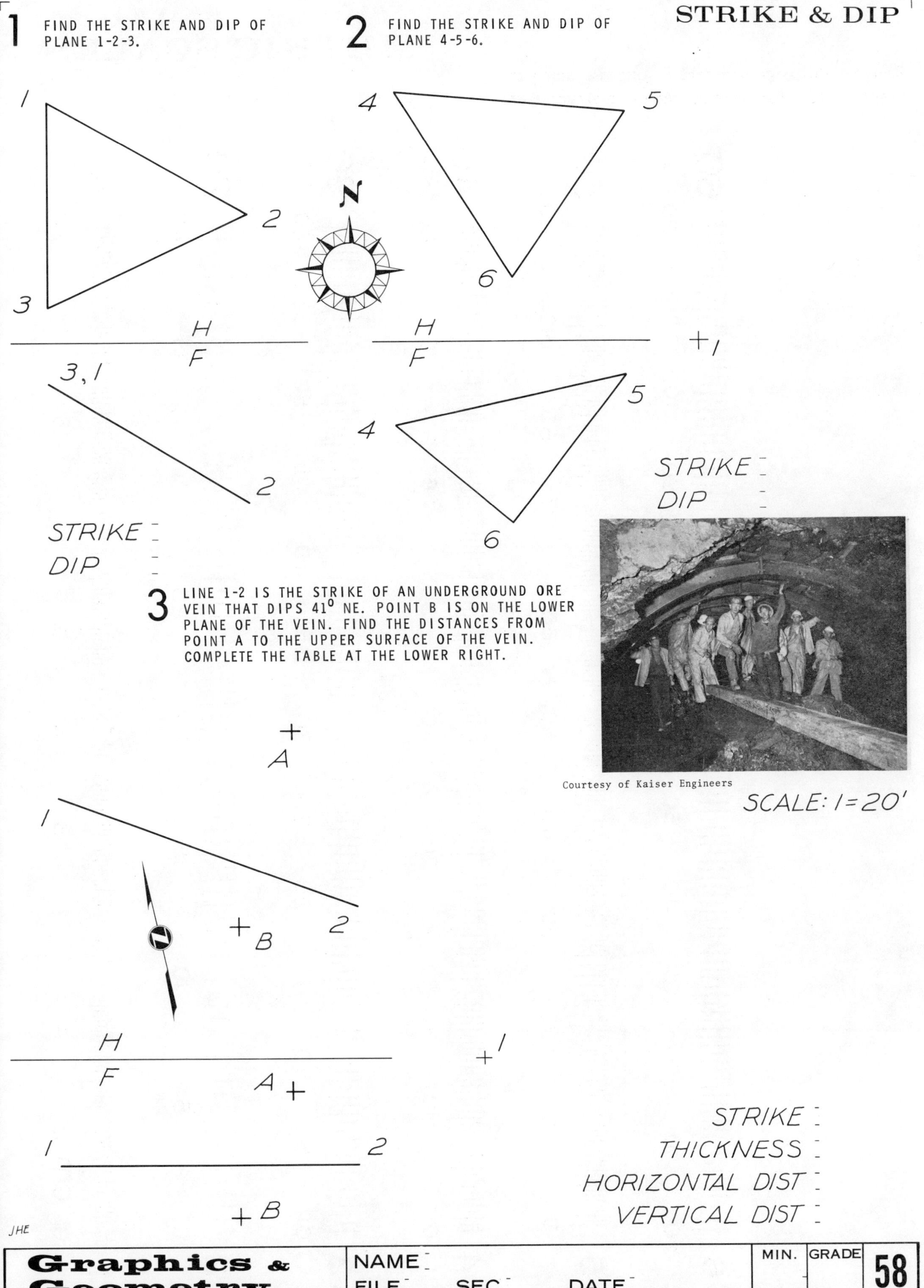

Courtesy of Kaiser Engineers

METRIC SCALES

Remove this page and fold it longwise along the desired scale for making metric measurements.

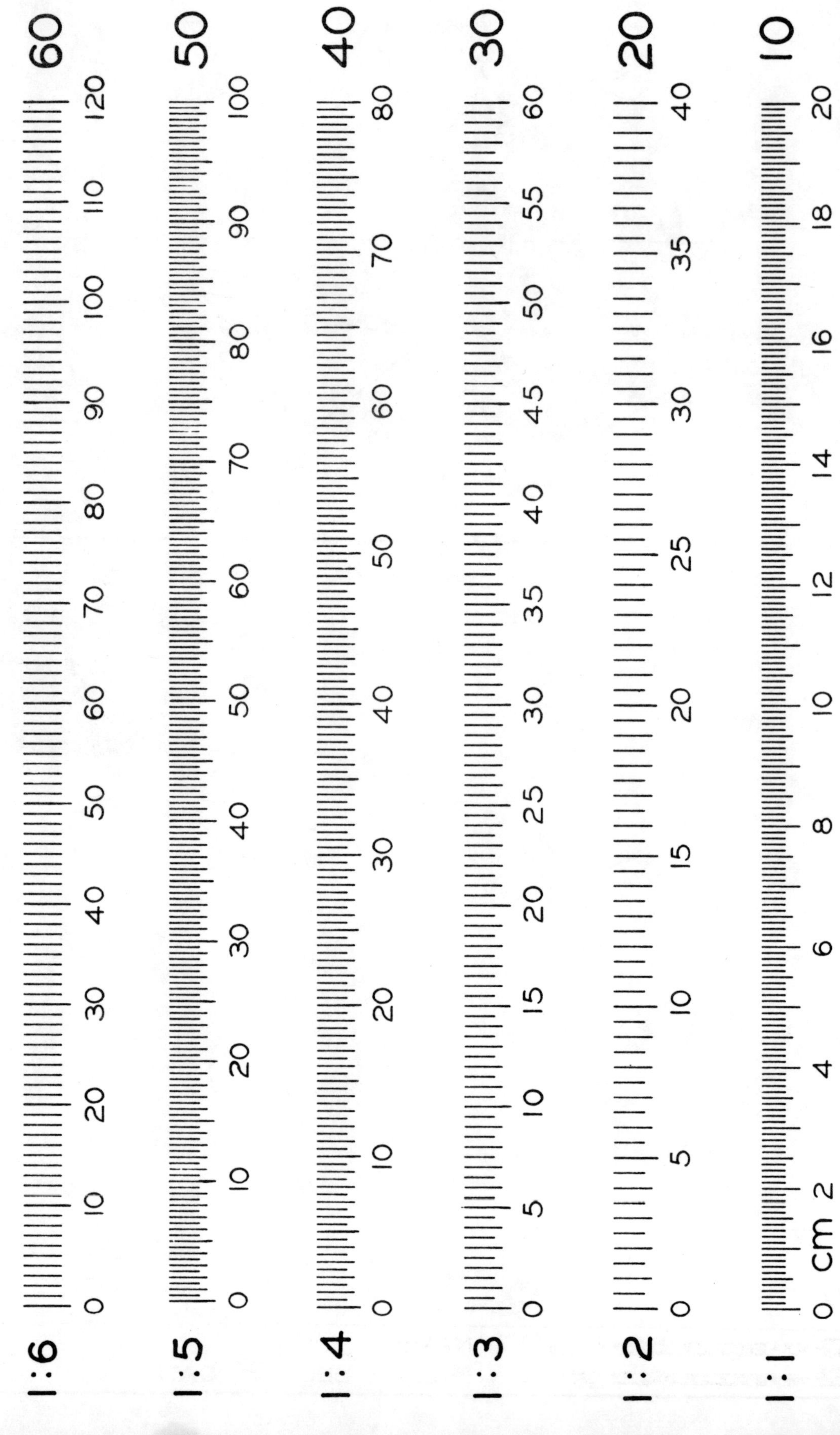

INTERSECTION OF PLANES

1 USE THE CUTTING PLANE METHOD TO FIND THE LINE OF INTERSECTION BETWEEN PLANES 1-2-3 AND 4-5-6. SHOW THE LINE IN BOTH VIEWS.

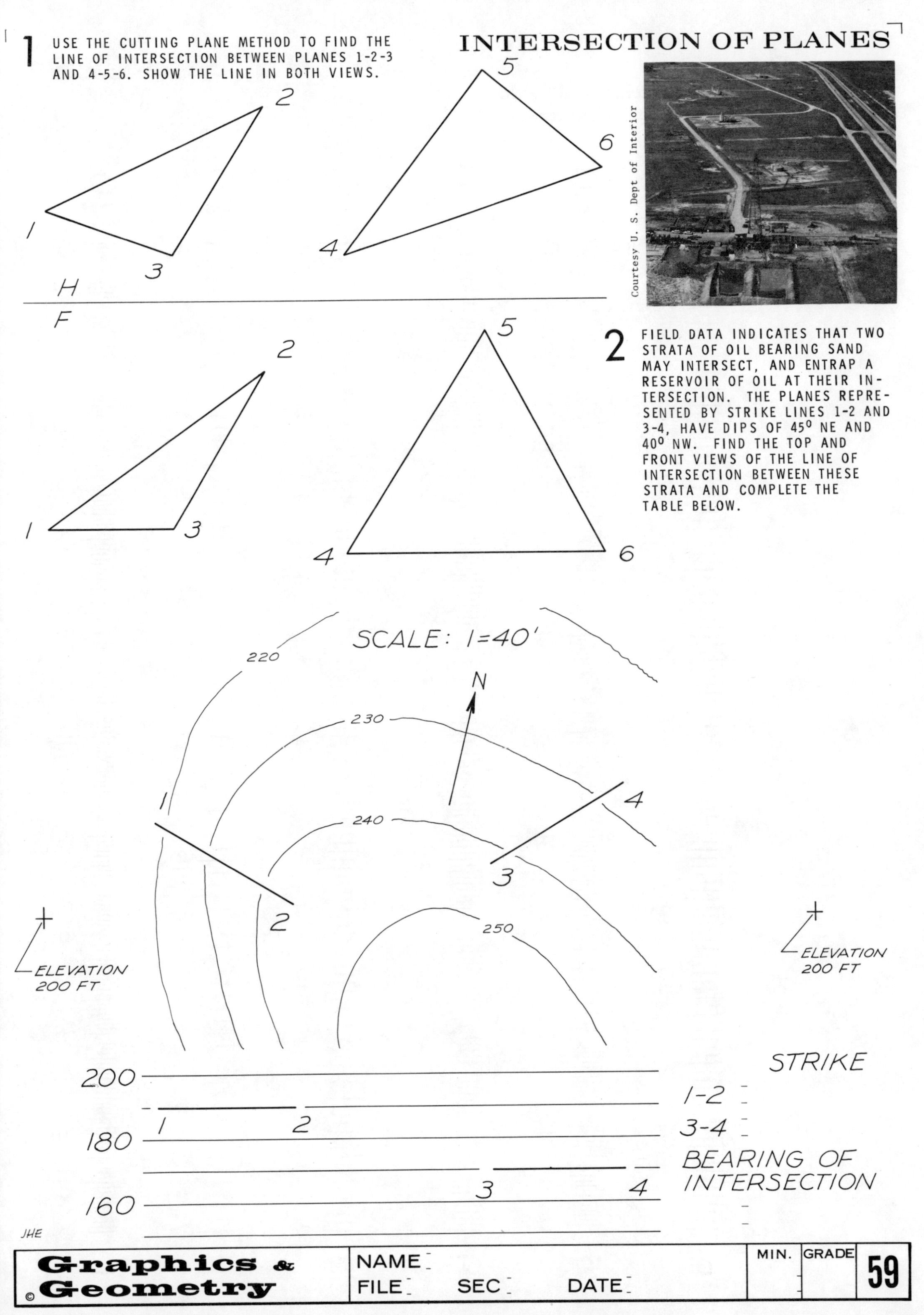

2 FIELD DATA INDICATES THAT TWO STRATA OF OIL BEARING SAND MAY INTERSECT, AND ENTRAP A RESERVOIR OF OIL AT THEIR INTERSECTION. THE PLANES REPRESENTED BY STRIKE LINES 1-2 AND 3-4, HAVE DIPS OF 45° NE AND 40° NW. FIND THE TOP AND FRONT VIEWS OF THE LINE OF INTERSECTION BETWEEN THESE STRATA AND COMPLETE THE TABLE BELOW.

STRIKE

1-2 ___

3-4 ___

BEARING OF INTERSECTION ___

METRIC SCALES

Remove this page and fold it longwise along the desired scale for making metric measurements.

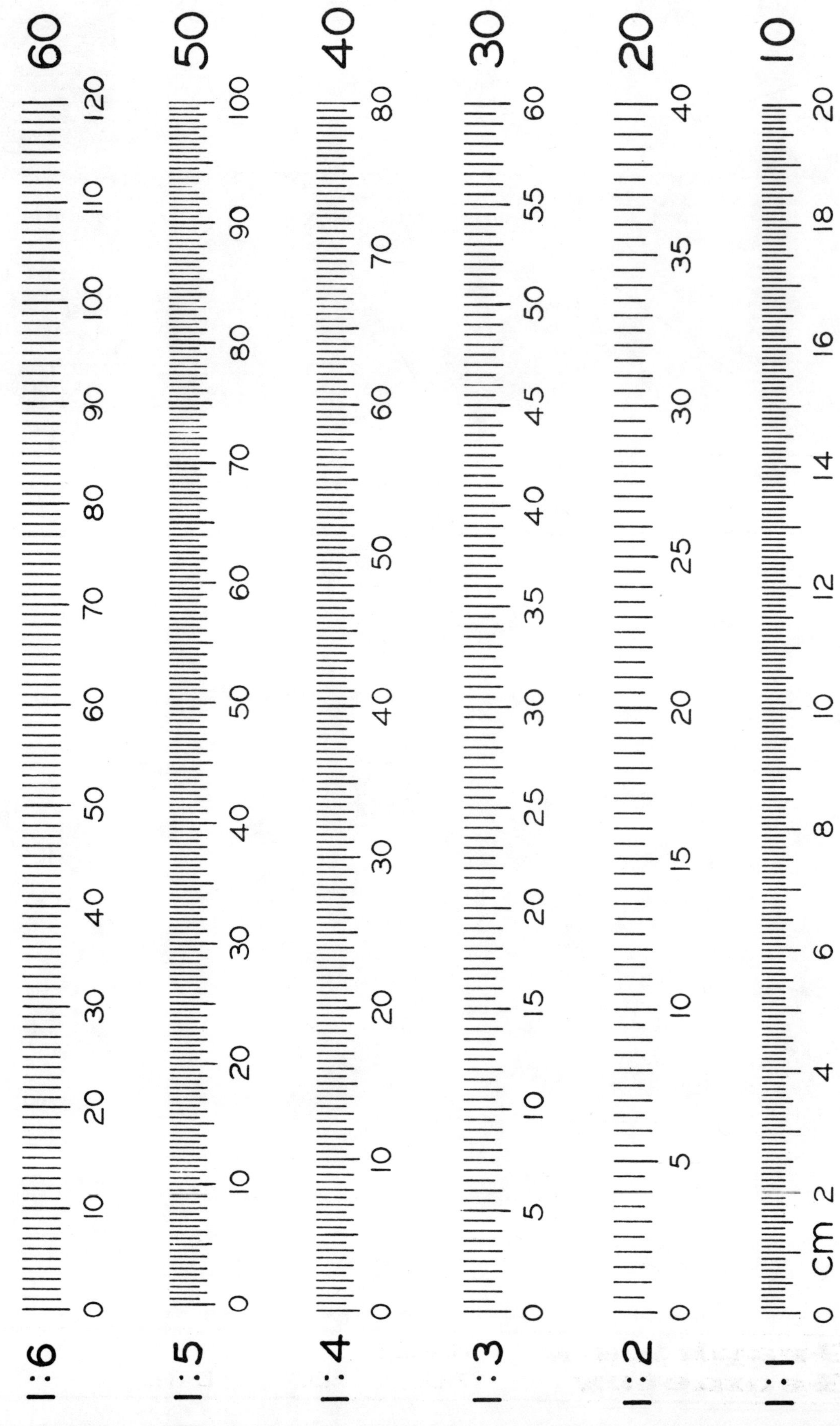

PLAN-PROFILE

IN ORDER TO SELL LOTS IN AN INDUSTRIAL TRACT THE DEVELOPER HAD TO PROVIDE SANITARY SEWERS FOR EACH PURCHASING INDUSTRY. FROM THE EXISTING 18 INCH SEWER LINE HE CONSTRUCTED A 12 INCH SEWER LINE FOLLOWING THE ROADS OF THE TRACT AS SHOWN IN THE PLAN BELOW. PLOT THE PROFILE OF THE GROUND AND THE 12 INCH SEWER LINE FROM MANHOLE #1 TO MANHOLE #3. THE FLOW LINE ELEVATION OF THE EXISTING SEWER AT MANHOLE #1 IS 224.50. USE A SCALE OF 1=10' FOR VERTICAL PLOTTING AND A SCALE OF 1=100' FOR HORIZONTAL PLOTTING. ASSUME ALL GRADES ARE POSITIVE FROM STATION 0+00 AND ALLOW 0.20 FEET OF FALL THROUGH EACH MANHOLE TO COMPENSATE FOR HEAD LOSS. WHERE THE SEWER LINE IS COVERED WITH LESS THAN 6 FEET OF SOIL IT MUST BE REINFORCED. DETAIL A SHOWS BOTH THE REINFORCING AND A TYPICAL MANHOLE.

HOW MANY FEET OF PIPE MUST BE REINFORCED? _

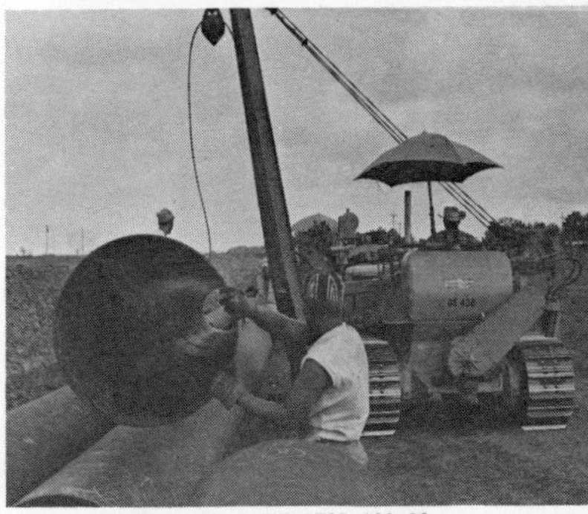

COURTESY LONE STAR GAS CO.

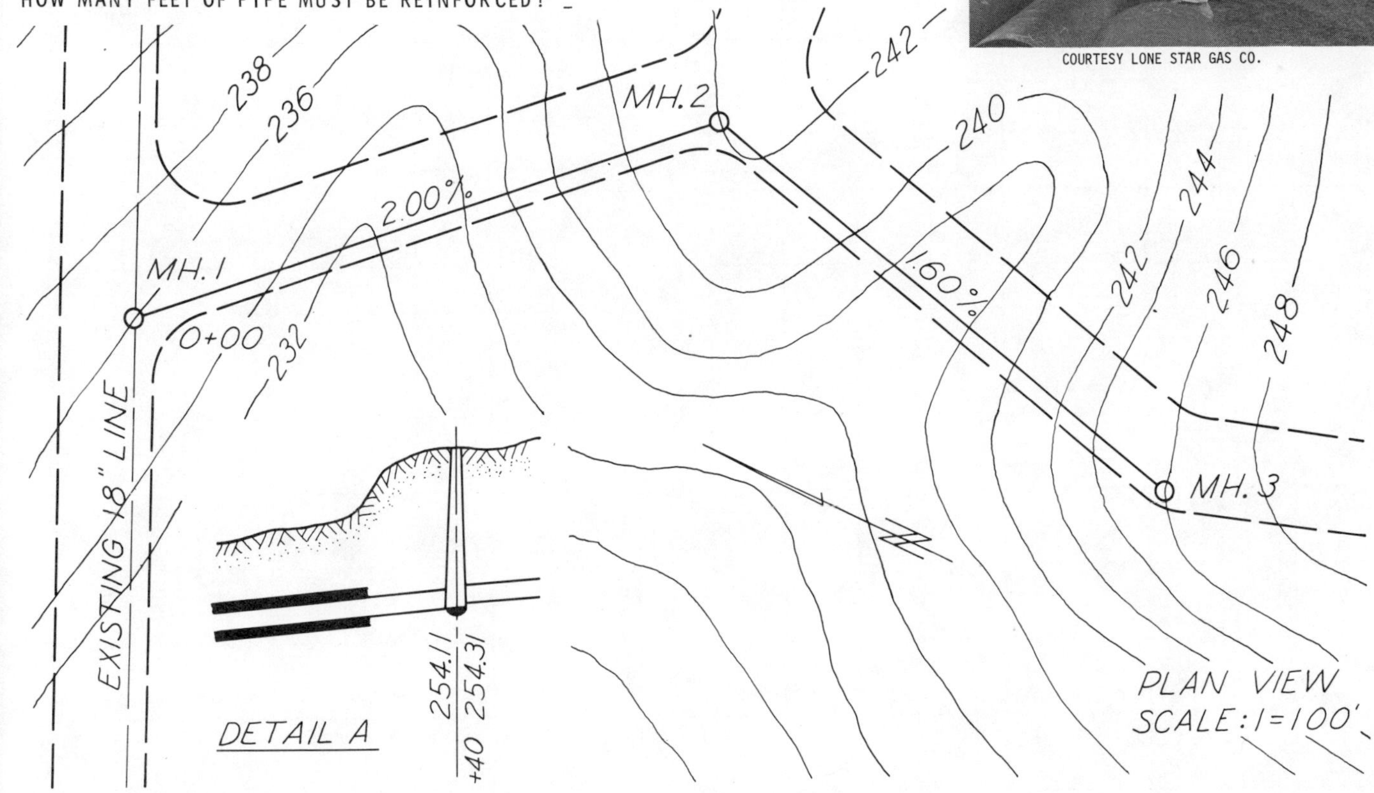

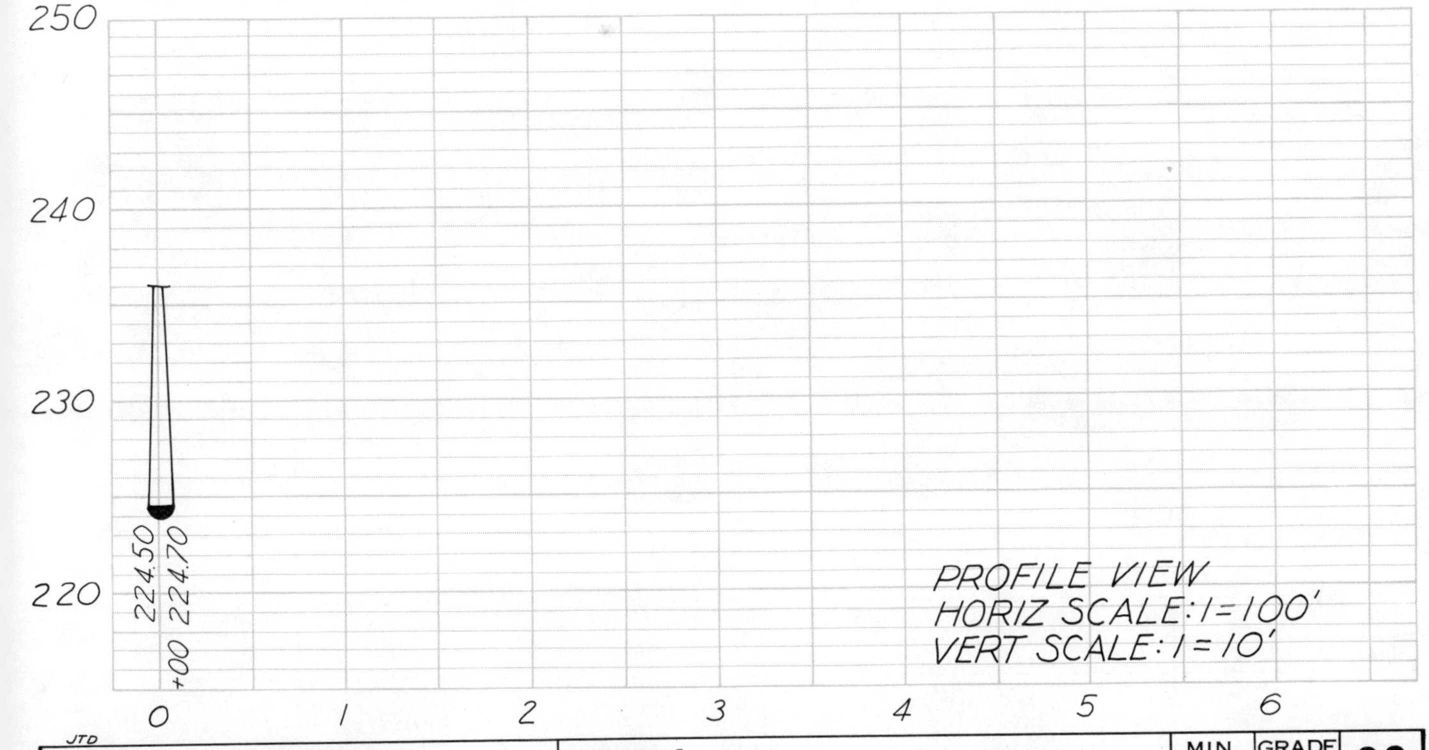

Graphics & Geometry	NAME _ FILE _ SEC _ DATE _	MIN.	GRADE	60

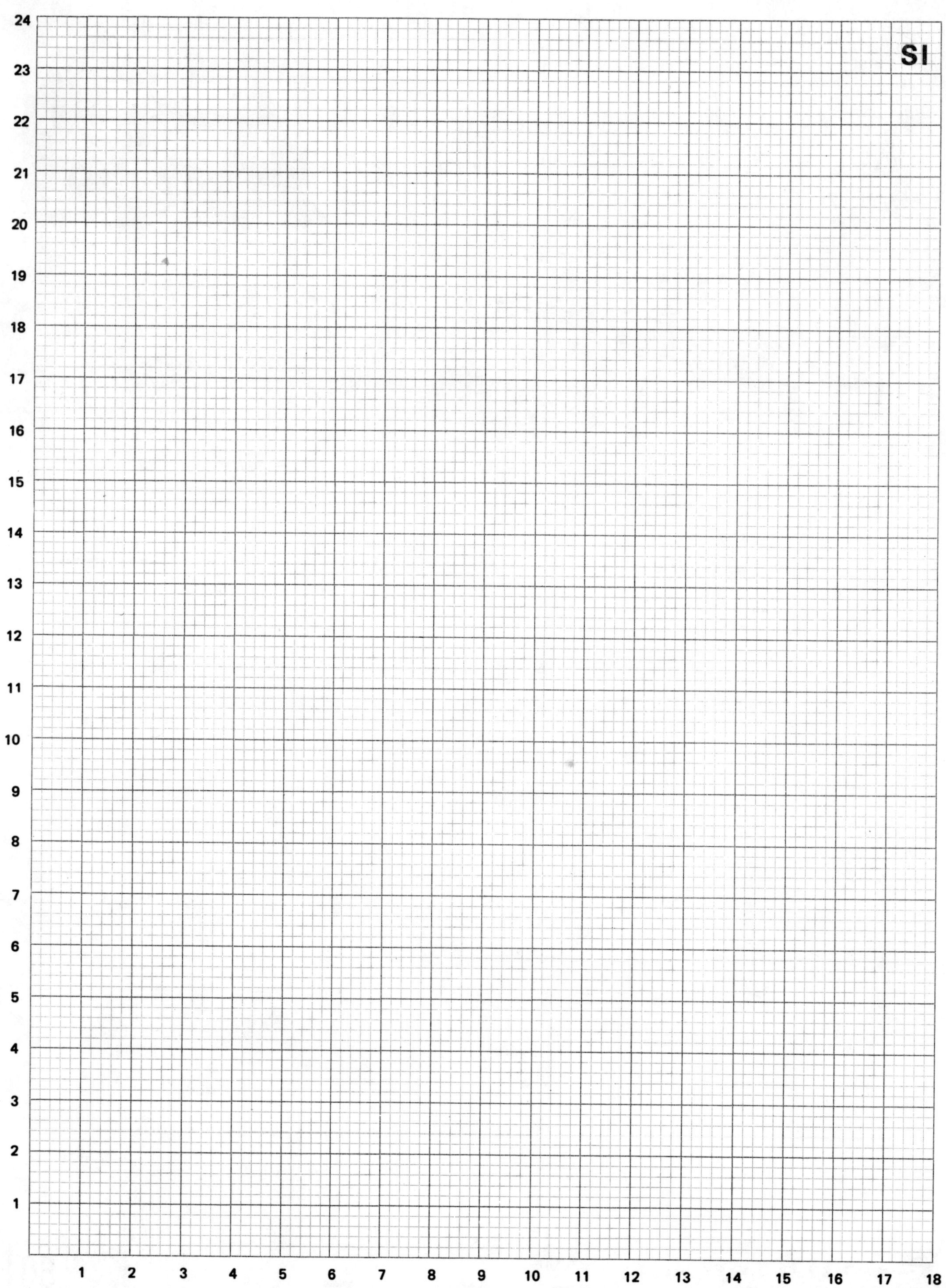
SI
24
23
22
21
20
19
18
17
16
15
14
13
12
11
10
9
8
7
6
5
4
3
2
1
1
2
3
4
5
6
7
8
9
10
11
12
13
14
15
16
17
18

ROADWAY CUT AND FILL

A LEVEL ROADWAY IS TO BE CONSTRUCTED AT ELEVATION 200 FEET WITH A WIDTH OF 60 FEET. THE CUT SLOPE IS 1:1 (THE RATIO OF RISE TO RUN) AND THE FILL SLOPE IS 1:1.7. PLOT THE VERTICAL SECTION TAKEN ALONG THE ROAD'S CUT AND FILL.

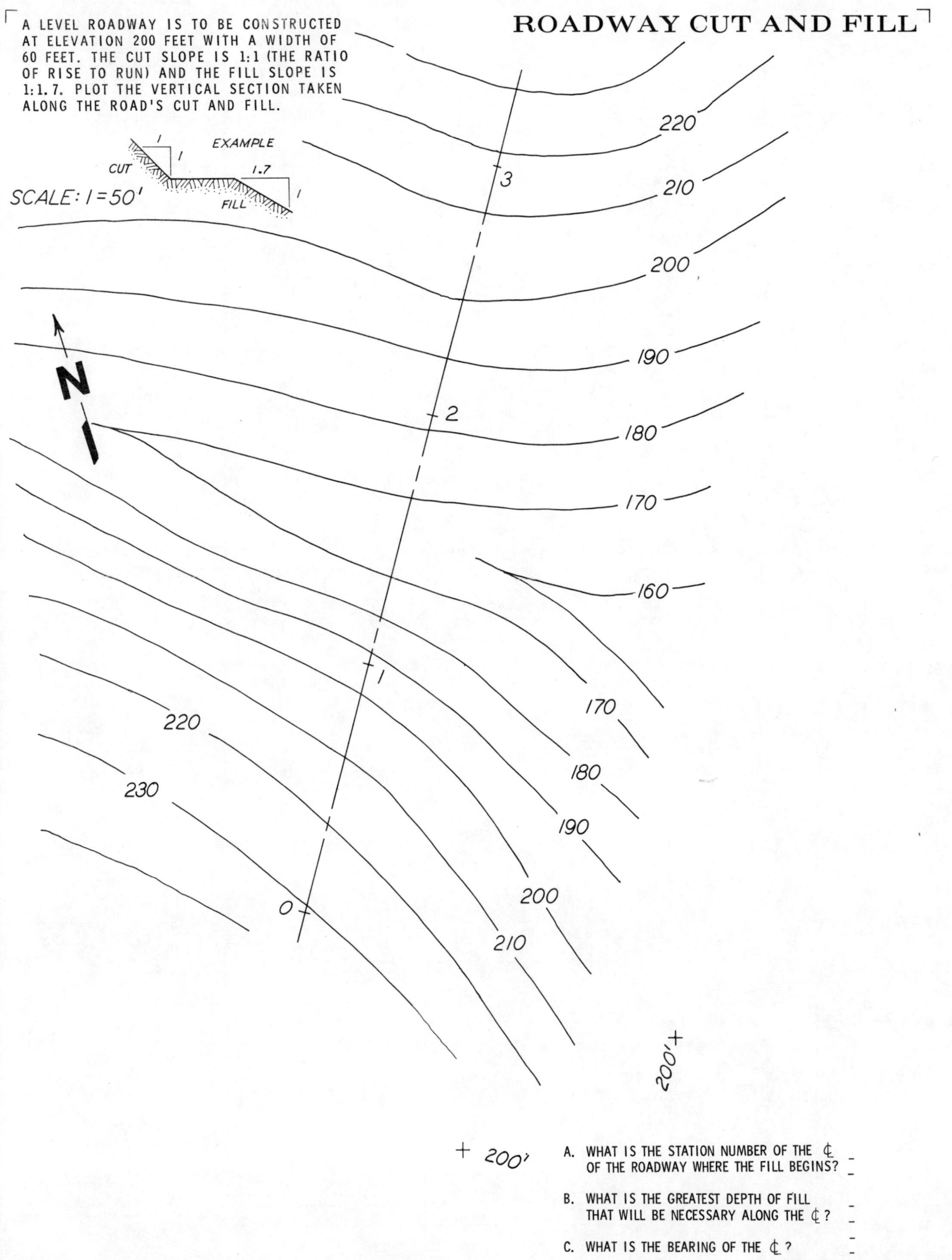

A. WHAT IS THE STATION NUMBER OF THE ℄ OF THE ROADWAY WHERE THE FILL BEGINS? _

B. WHAT IS THE GREATEST DEPTH OF FILL THAT WILL BE NECESSARY ALONG THE ℄? _

C. WHAT IS THE BEARING OF THE ℄? _

SI
24
23
22
21
20
19
18
17
16
15
14
13
12
11
10
9
8
7
6
5
4
3
2
1
1 2 3 4 5 6 7 8 9 10 11 12 13 14 15 16 17 18

EARTHEN DAM

FIND THE LIMITS OF FILL FOR A CURVED EARTHEN DAM THAT MEETS THE FOLLOWING SPECIFICATIONS; AND CONSTRUCT A VERTICAL SECTION TAKEN ALONG THE CENTER LINE.

<u>SPECIFICATIONS:</u>

CREST WIDTH--10 FT
CREST ELEVATION--340 FT
CREST RADIUS--70 FT FROM CENTER C
UPSTREAM SLOPE--1:1.5
DOWNSTREAM SLOPE--1:1
WATER ELEVATION--333 FT

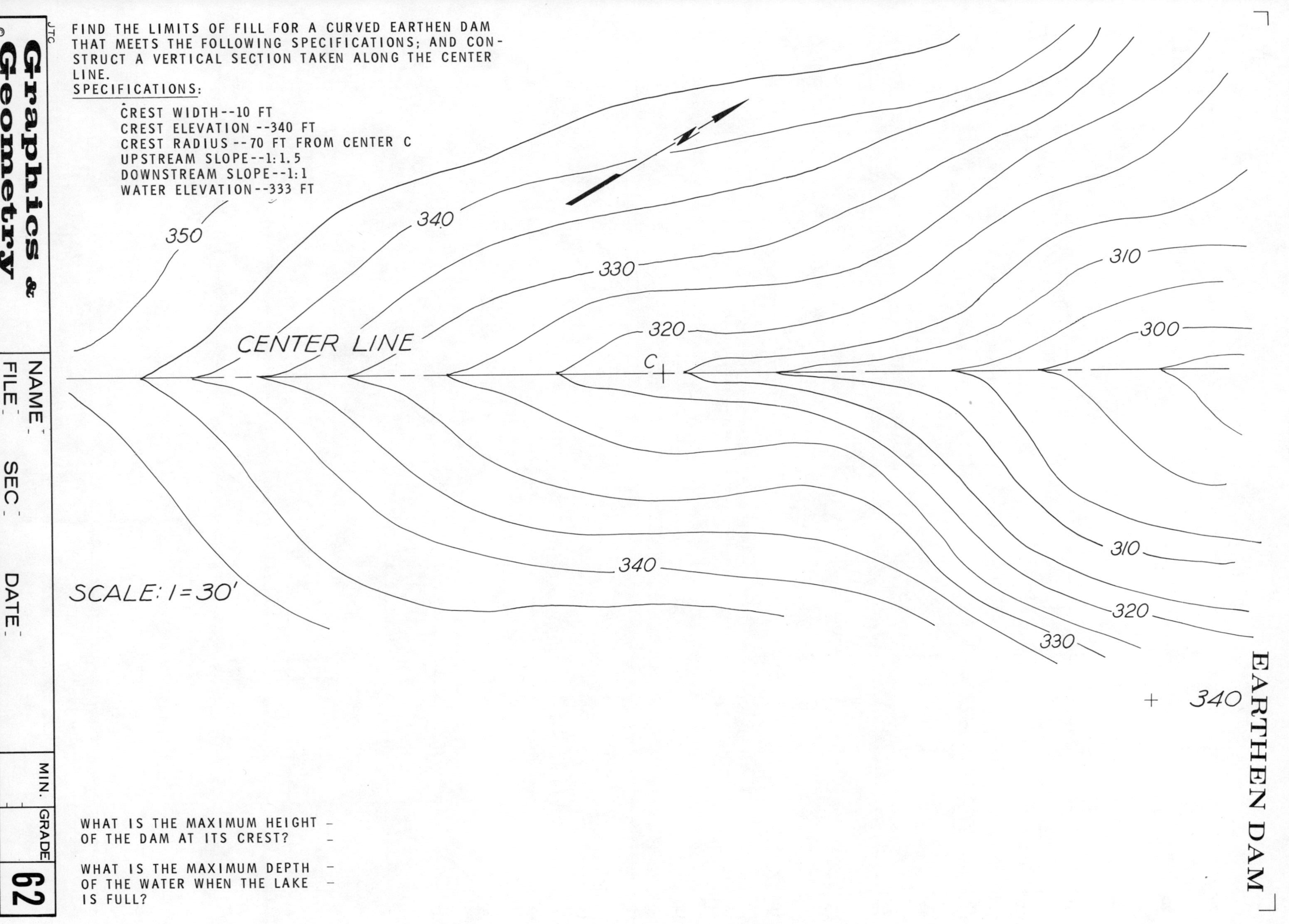

WHAT IS THE MAXIMUM HEIGHT OF THE DAM AT ITS CREST? ___

WHAT IS THE MAXIMUM DEPTH OF THE WATER WHEN THE LAKE IS FULL? ___

JTC

Graphics & Geometry ©	NAME ___ FILE ___ SEC ___ DATE ___	MIN.	GRADE	62

SI
24
23
22
21
20
19
18
17
16
15
14
13
12
11
10
9
8
7
6
5
4
3
2
1
1 2 3 4 5 6 7 8 9 10 11 12 13 14 15 16 17 18

OUTCROP

POINT A IS LOCATED ON THE TOP OF AN ORE VEIN AT AN ELEVATION OF 75 FT. THE VEIN HAS A STRIKE OF S 60° E, A DIP OF 30° SW, AND A THICKNESS OF 30 FT.

1 FIND THE OUTCROP OF THE ORE VEIN ON THE SURFACE OF THE GROUND. CONSTRUCT THE PROFILE VIEW SHOWING ELEVATION LINES, THE ORE VEIN AND A VERTICAL SECTION OF THE GROUND WHICH PASSES THROUGH POINT B.

2 FROM POINT B, ON THE SURFACE OF THE GROUND, FIND THE FOLLOWING DISTANCES TO THE TOP OF THE ORE VEIN:
(A) THE SHORTEST DISTANCE TO THE ORE VEIN,
(B) THE VERTICAL DISTANCE TO THE ORE VEIN.

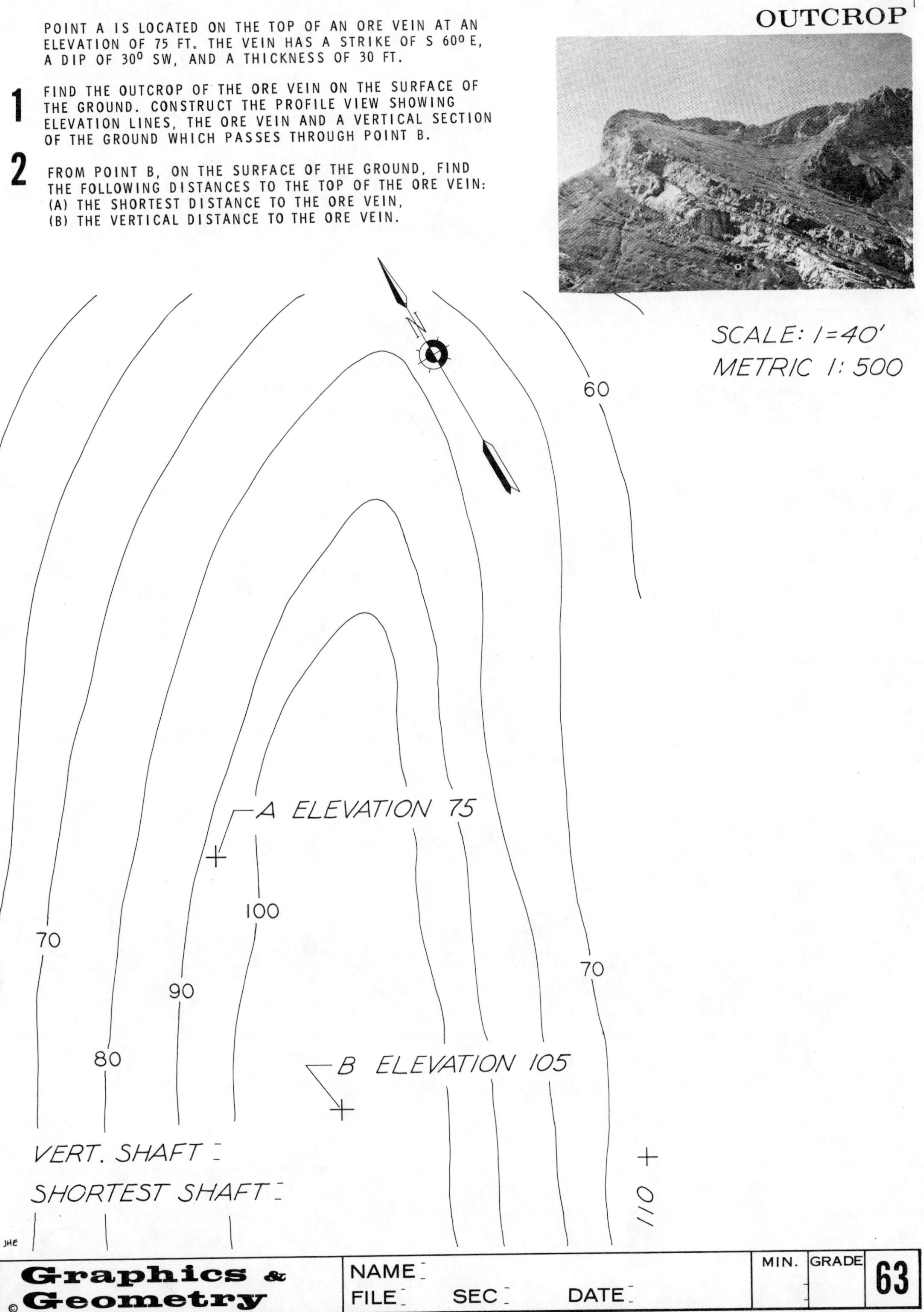

VERT. SHAFT ___

SHORTEST SHAFT ___

JHE

Graphics & Geometry	NAME ___ FILE ___ SEC ___ DATE ___	MIN.	GRADE	63

SI

SUCCESSIVE AUXILIARY VIEWS

PROBLEM 1. DRAW PARTIAL VIEWS OF THE OBLIQUE CLEVIS THAT WILL SHOW THE OBLIQUE SURFACE TRUE SIZE.

PROBLEM 2. COMPLETE ALL VIEWS OF THE OBLIQUE CLEVIS BY USING ELLIPSE GUIDES.

NOTE: A. THE CLEVIS IS MADE OF 5/8" PLATE

B. FILLETS 1/4" R.

1

2

H

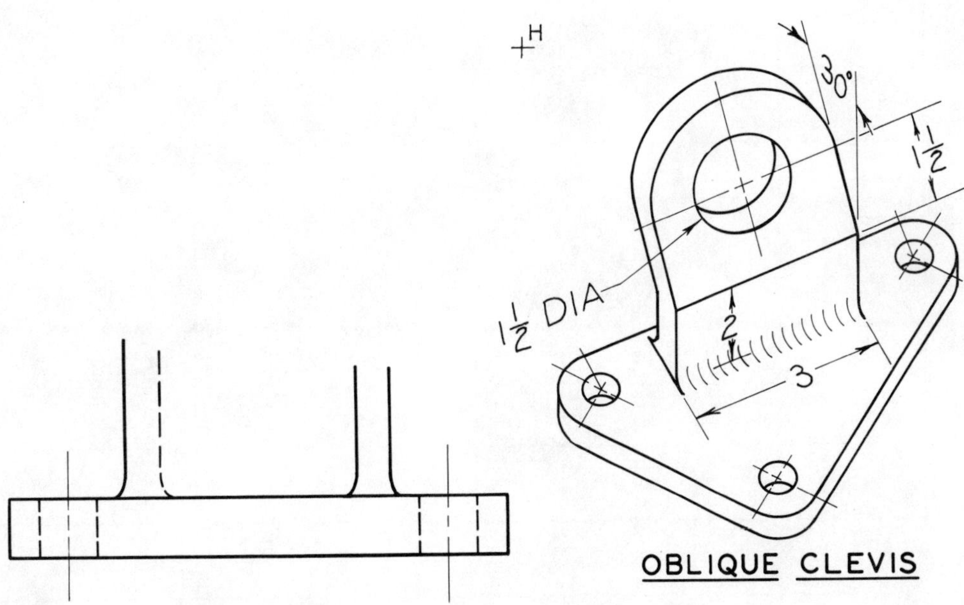

OBLIQUE CLEVIS

HALF SIZE

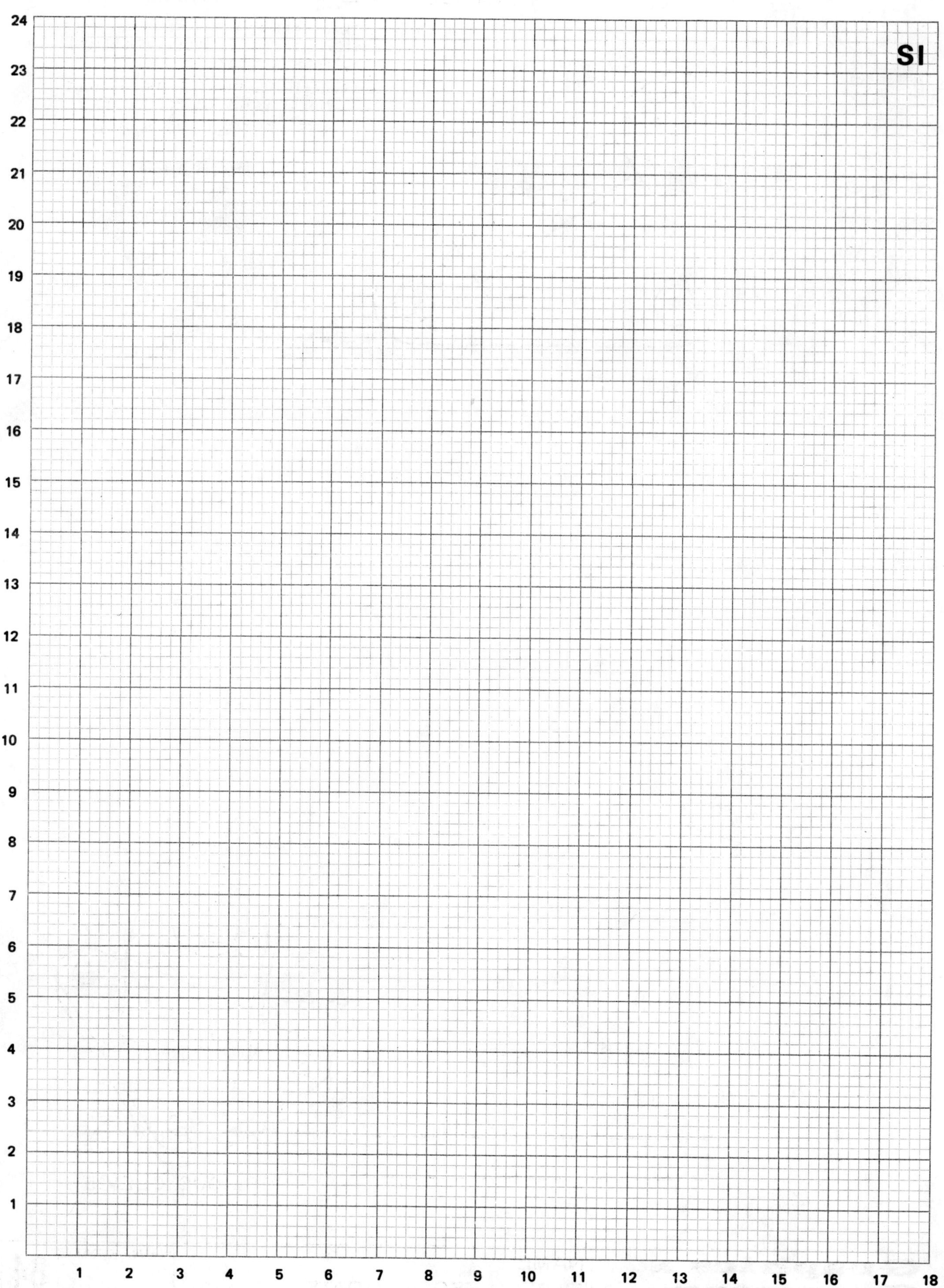
SI
24
23
22
21
20
19
18
17
16
15
14
13
12
11
10
9
8
7
6
5
4
3
2
1
1 2 3 4 5 6 7 8 9 10 11 12 13 14 15 16 17 18

1 DRAW THE VIEW THAT SHOWS THE TRUE ANGLE BETWEEN LINES 1-2 AND 2-3.

THE ANGLE IS:

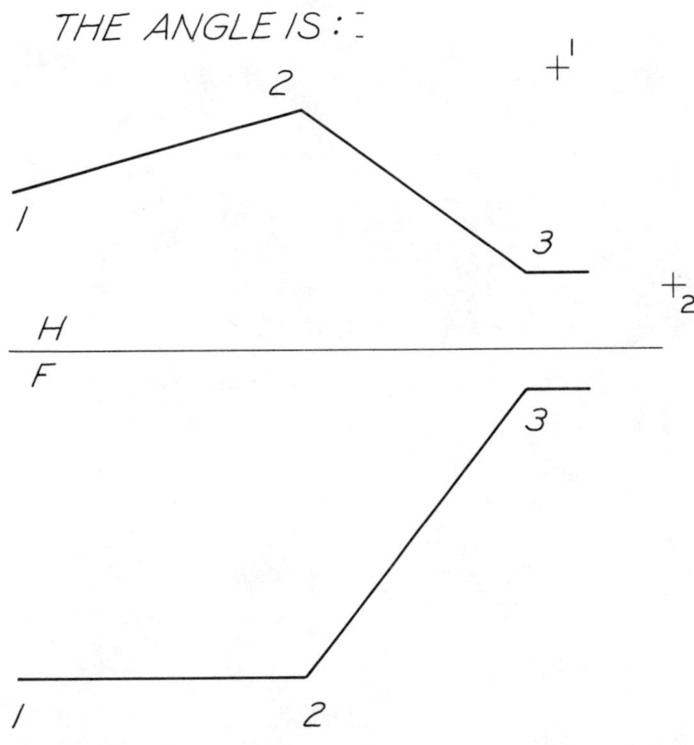

2 IN THE CONSTRUCTION OF PIPE LINES IT IS NECESSARY TO DETERMINE THE TRUE ANGLES OF BENDS IN PIPES. CONSTRUCT A BEND OF 3' (1 m) RADIUS IN THE VIEW THAT SHOWS THIS ANGLE.

THE ANGLE IS

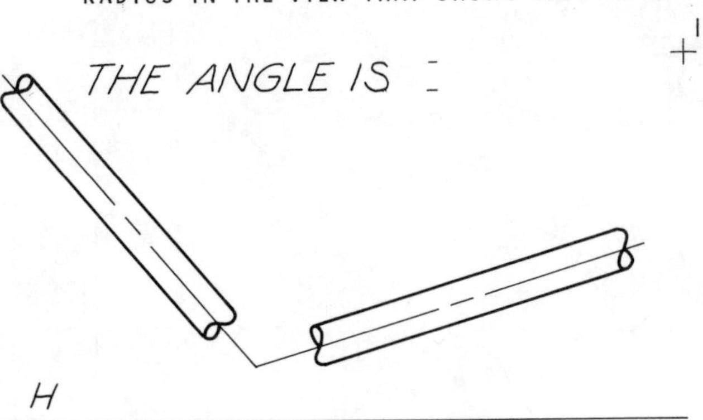

SCALE: 1 = 4'
1:50

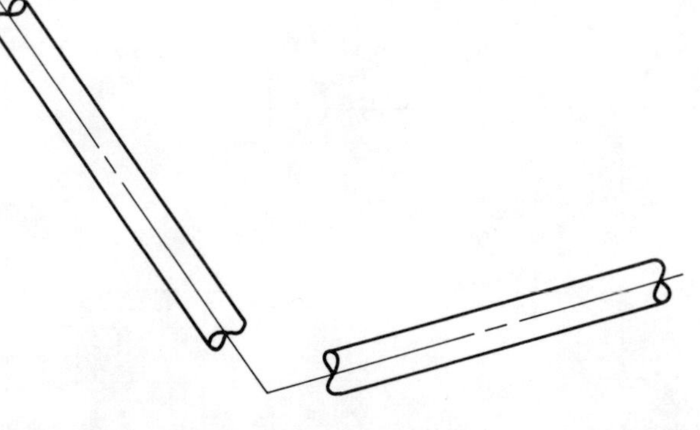

COURTSEY PHILLIPS PETROLEUM CO.

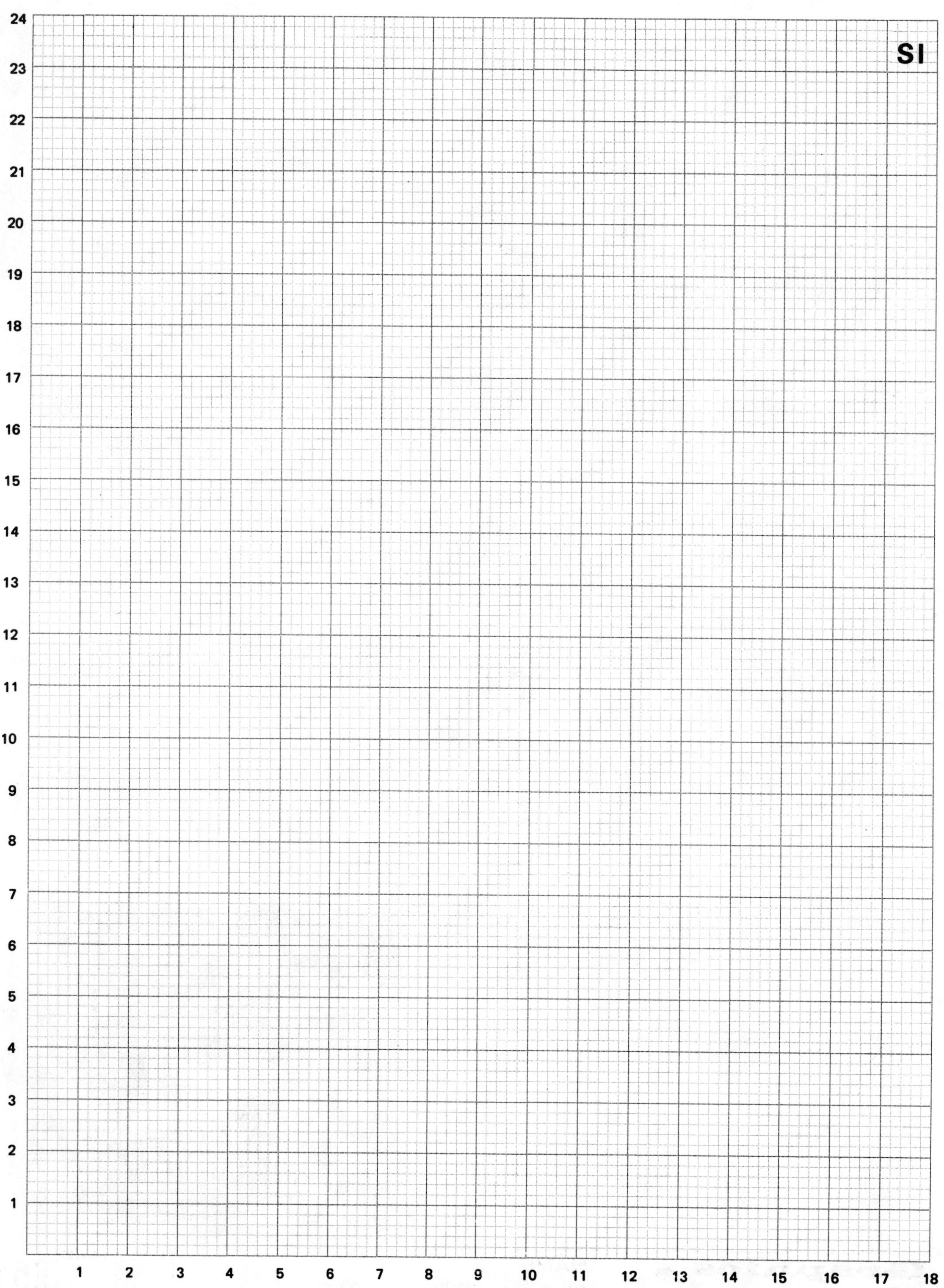
SI

ANGLE BETWEEN PLANES

1 DRAW THE VIEW THAT SHOWS THE TRUE ANGLE BETWEEN PLANES 1-2-3 AND 1-2-4.

ANGLE =

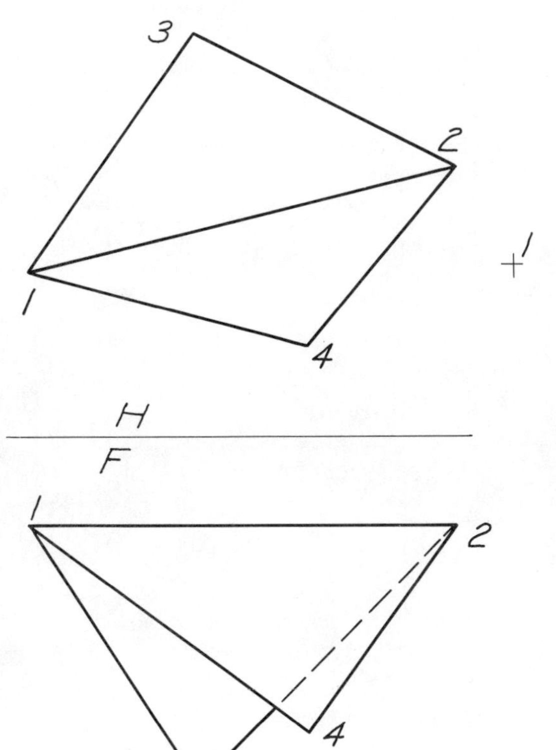

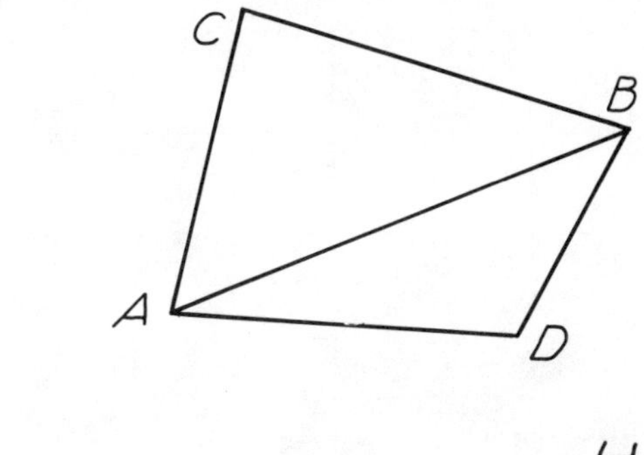

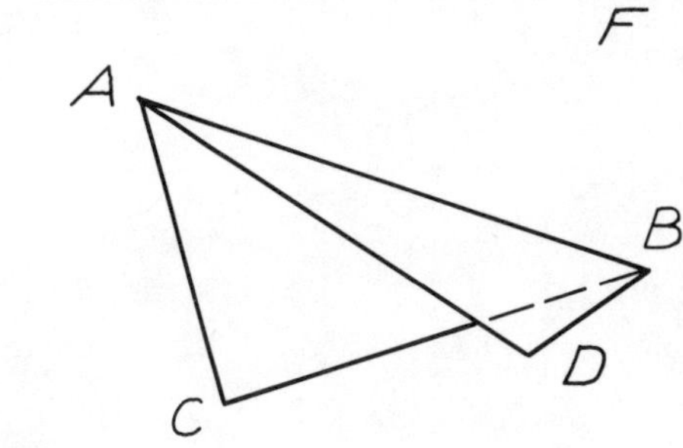

ANGLE =

2 DRAW THE VIEW THAT SHOWS THE TRUE ANGLE BETWEEN PLANES ABC AND BCD.

3 IN ORDER TO BUILD A DIFFUSER SECTION SIMILAR TO THE ONE IN THE PHOTOGRAPH, THE DIHEDRAL ANGLE MUST BE KNOWN. USE AB AS THE LINE OF INTERSECTION IN QUESTION.

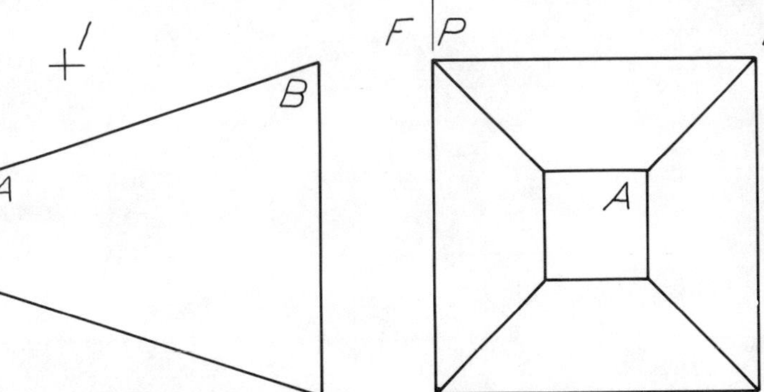

ANGLE =

SI
24
23
22
21
20
19
18
17
16
15
14
13
12
11
10
9
8
7
6
5
4
3
2
1
1 2 3 4 5 6 7 8 9 10 11 12 13 14 15 16 17 18

DISTANCE POINT TO LINE

1 FIND THE SHORTEST DISTANCE FROM POINT 3 TO LINE 1-2 BY THE LINE METHOD. SHOW IN ALL VIEWS.

DISTANCE _

DISTANCE _

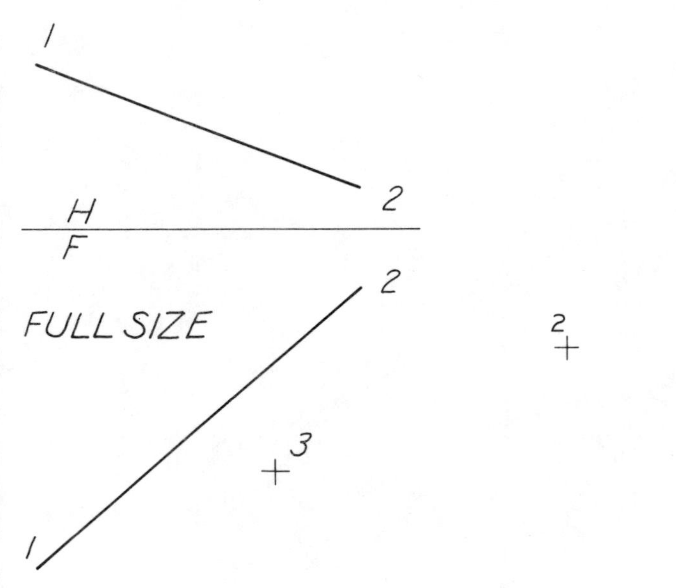

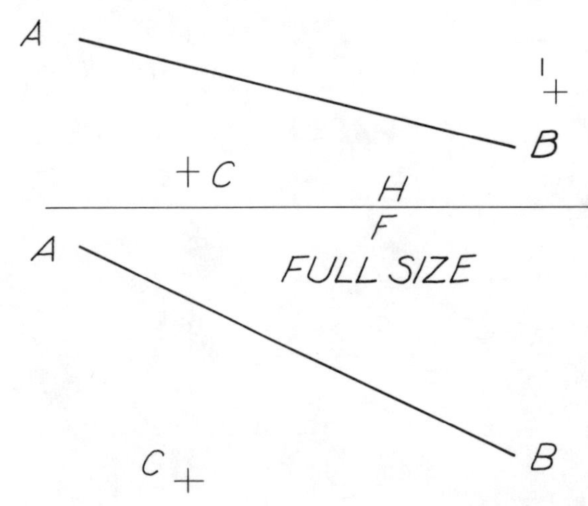

2 FIND THE SHORTEST DISTANCE FROM POINT C TO LINE AB AND SHOW IT IN ALL VIEWS.

3 IN THE ROUTING OF BUS BARS THROUGH POWER SUBSTATIONS, IT IS NECESSARY TO MAINTAIN CERTAIN CLEARANCES BETWEEN THE BUS BARS AND THE SUPPORTING STRUCTURES. HOW FAR CAN THE VERTICAL POST CENTER LINE PO BE EXTENDED AND STILL MAINTAIN A CLEARANCE OF 1.5' OR 0.46 m? SHOW EXTENSION OF PO IN ALL VIEWS.

PO MAY BE EXTENDED _

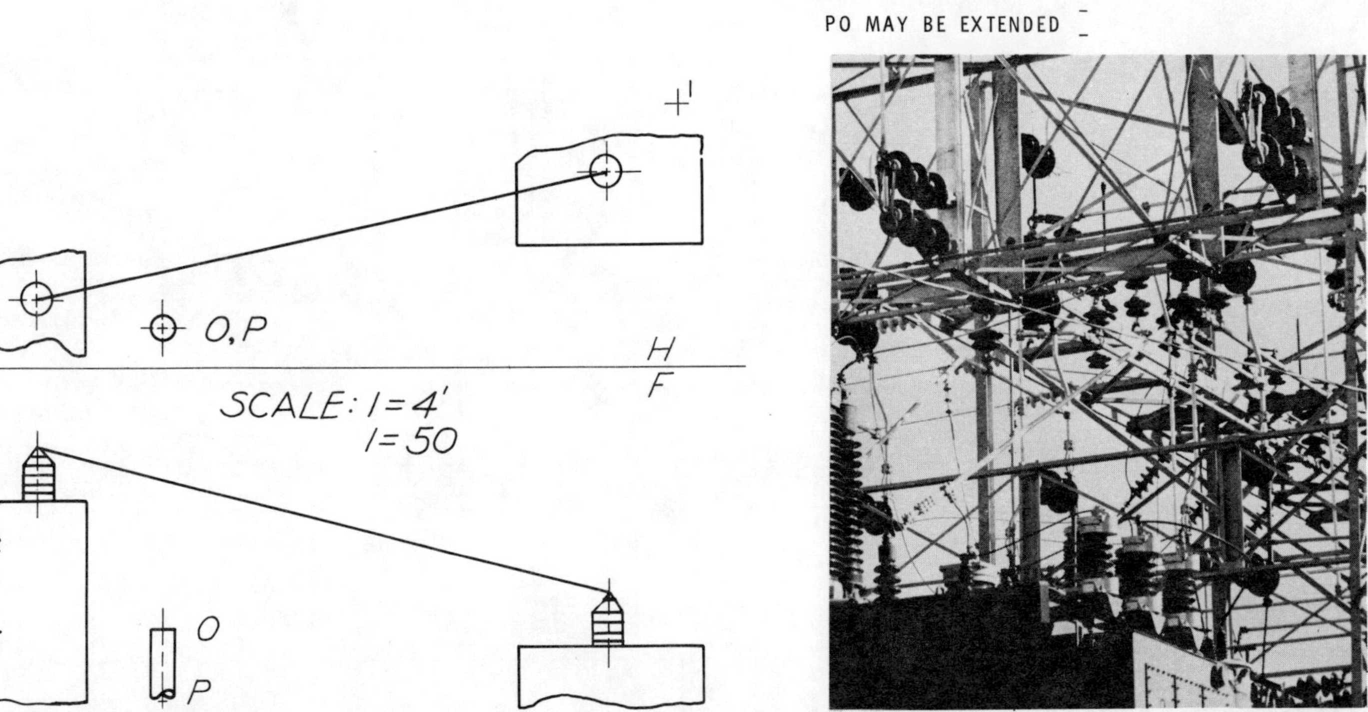

JTC

SI
24
23
22
21
20
19
18
17
16
15
14
13
12
11
10
9
8
7
6
5
4
3
2
1
1 2 3 4 5 6 7 8 9 10 11 12 13 14 15 16 17 18

ANGULAR DISTANCE

1 CONSTRUCT A LINE FROM POINT O IN A DOWNWARD DIRECTION THAT WILL MAKE AN ANGLE OF 45° WITH LINE 1-2. SHOW IN ALL VIEWS.

DISTANCE:

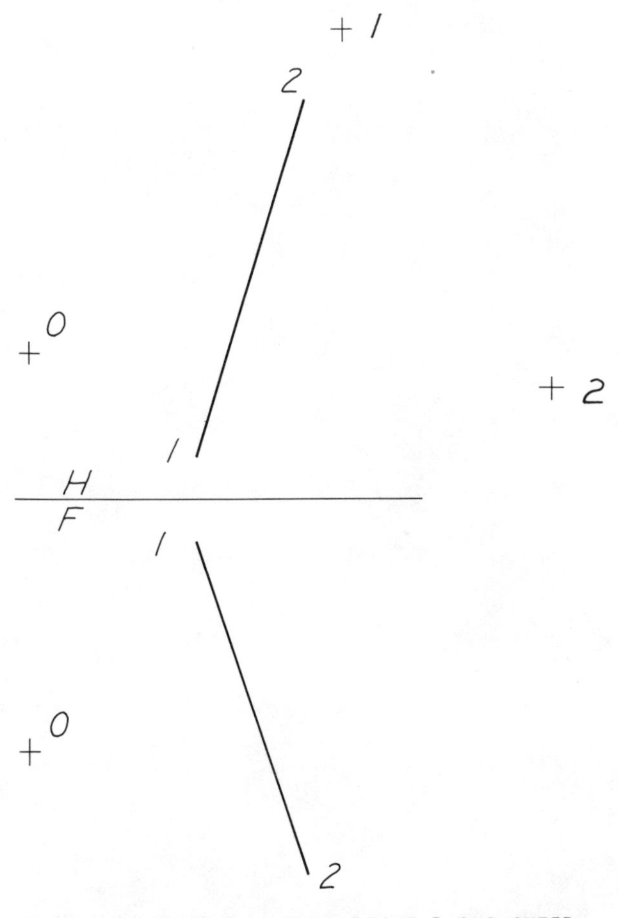

2
+

+ 1

2 A PIPE IS TO RUN FROM POINT P AND INTERSECT THE PIPE AB AT A 45° ANGLE. CONSTRUCT THIS PIPE AND SHOW IT IN ALL VIEWS. DRAW THE 125 LB CAST IRON PIPE LATERAL IN ITS TRUE SIZE VIEW USING DOUBLE-LINE SYMBOLS.

B

A

+ P

H

F

A

B

SCALE: 1=10'
1:100

P
+

DISTANCE:

COURTSEY STEARNS-ROGER CORP.

Graphics & Geometry ©	NAME FILE SEC DATE	MIN.	GRADE	68

SKEWED LINES

1 FIND THE SHORTEST DISTANCE BETWEEN THE TWO LINES BY THE LINE METHOD. SHOW THE LINE IN ALL VIEWS.

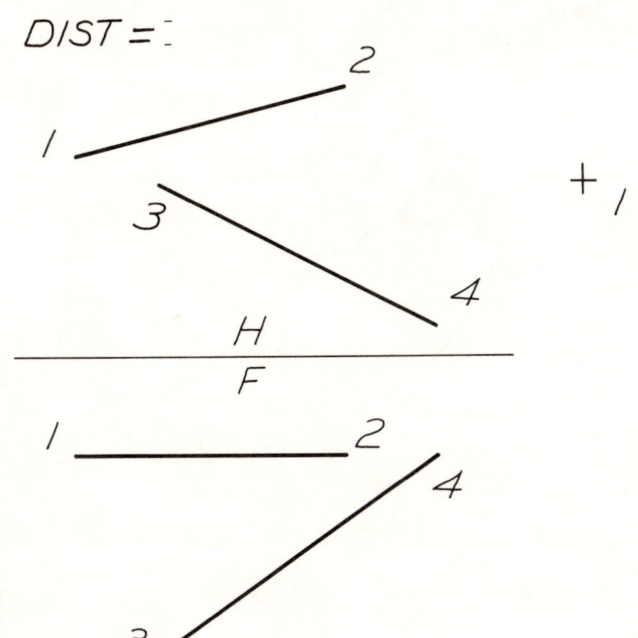

Courtesy U. S. Dept of Interior

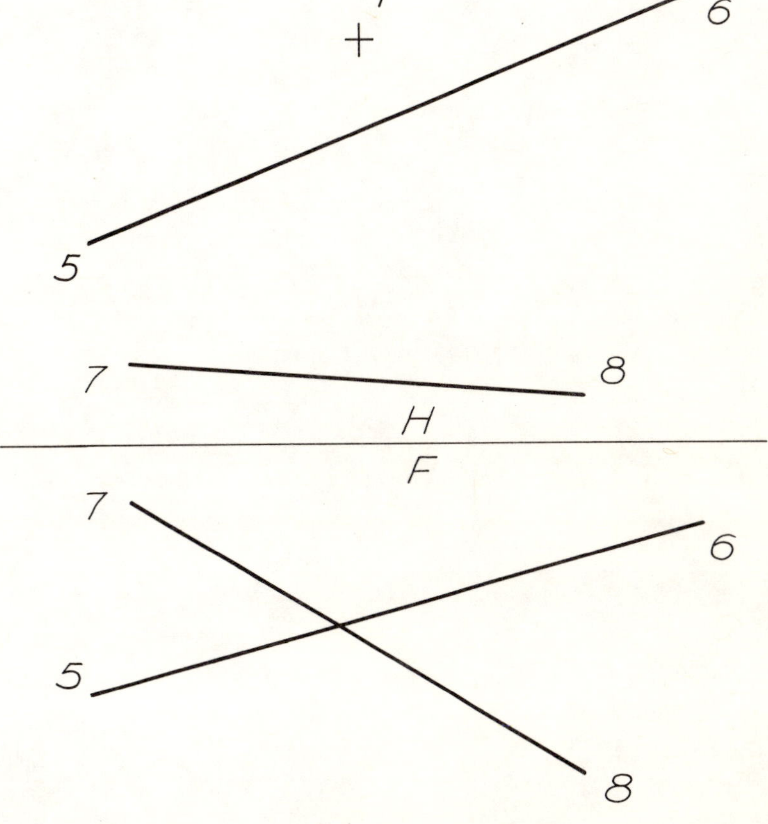

2 +

SCALE: 1 = 5'
1:60 mm

SHORTEST DIST

2

POWER LINES THAT CARRY HIGH VOLTAGES MUST BE SEPARATED BY A SUFFICIENT DISTANCE TO PREVENT INDUCTION DUE TO INTERFERING MAGNETIC FIELDS.

DETERMINE THE SHORTEST DISTANCE BETWEEN POWER LINES 5-6 AND 7-8 BY THE LINE METHOD. BEGIN BY FINDING LINE 5-6 TRUE LENGTH BY PROJECTING FROM THE TOP VIEW. SHOW THE SHORTEST DISTANCE IN ALL VIEWS.

JHE

HIGHWAY CONSTRUCTION COSTS INVOLVE ENGINEERING DESIGN; RIGHT-OF-WAY PURCHASE; PREPARATION OF SUBGRADE; BASE AND RIDING SURFACE; MATERIALS AND LABOR. BECAUSE OF THESE COSTS IN DESIGNING A HIGHWAY FROM ONE POINT TO ANOTHER, THE SHORTEST DISTANCE IS THE MOST DESIRABLE, BUT TO REACH A MOUNTAIN TOP BY AUTOMOBILE, A HIGHWAY DESIGNER DOES NOT TAKE THE SHORTEST DISTANCE TO THE TOP. HE DESIGNS TO PERCENT GRADES THAT WILL ENABLE AN AUTOMOBILE TO MAINTAIN A REASONABLE SPEED WITHOUT DANGER OF OVERHEATING OR UNDUE STRAIN. SHOW DISTANCES IN ALL VIEWS.

1. FIND THE SHORTEST 25 PERCENT GRADE CENTER LINE BETWEEN THE TWO CENTER LINES OF STATE HIGHWAYS 6 AND 21.
2. WHAT IS THE MAXIMUM VEHICLE HEIGHT THAT CAN BE DRIVEN THROUGH THE UNDERPASS AT THE CENTER LINES? ALLOW A VERTICAL CLEARANCE OF TWO FEET.

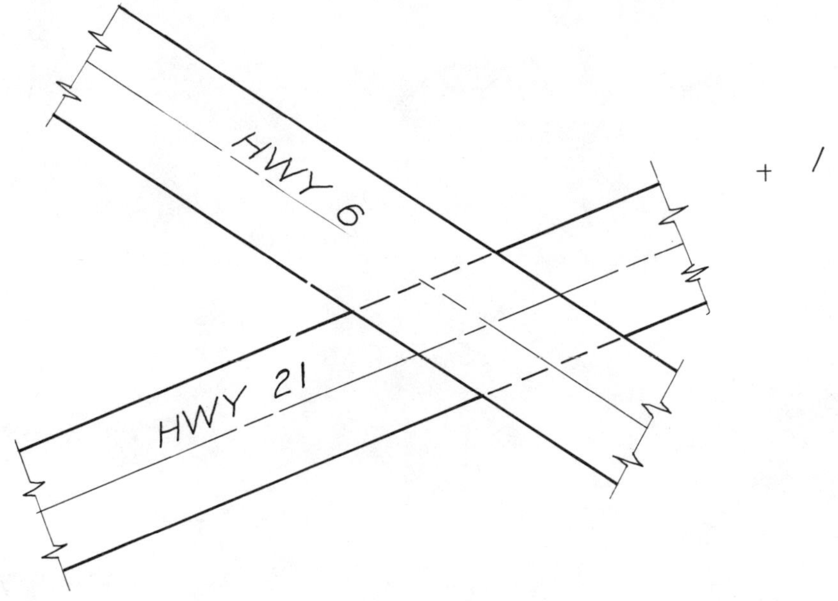

1

2

H

F

HWY 6

HWY 21

SCALE: 1 = 40'

METRIC 1:500

25% GRADE DIST =

MAX VEHICLE HT =

COURTESY TEXAS HIGHWAY DEPARTMENT

LES

ANGLE:

ANGLE BETWEEN LINE & PLANE

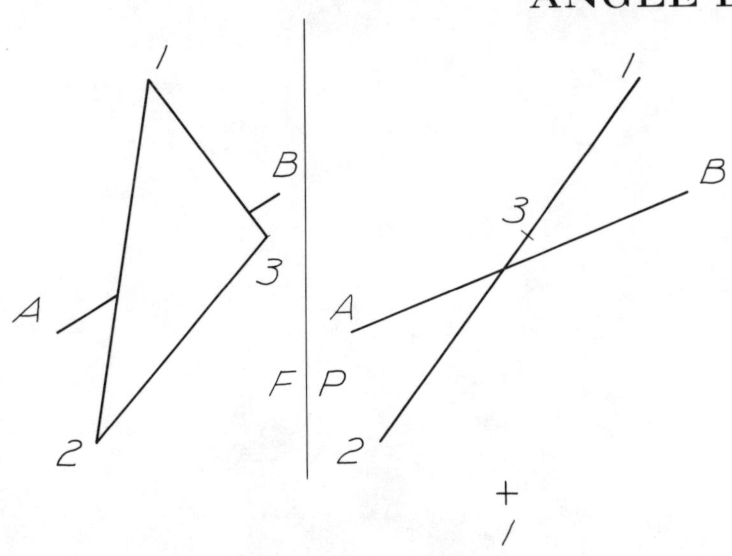

1 FIND THE ANGLE BETWEEN THE LINE AND THE PLANE BY THE PLANE METHOD. SHOW VISIBILITY IN ALL VIEWS.

2

AN ASTRONAUT'S LINE OF SIGHT IS ALONG LINE DC WHICH INTERSECTS THE TRIANGULAR WINDOW OF A SPACECRAFT.

DETERMINE THE ANGLE BETWEEN THE LINE AND PLANE BY THE PLANE METHOD. PROJECT FROM THE TOP VIEW. SHOW VISIBILITY IN ALL VIEWS.

Courtesy of Ryan Aeronautical Co.

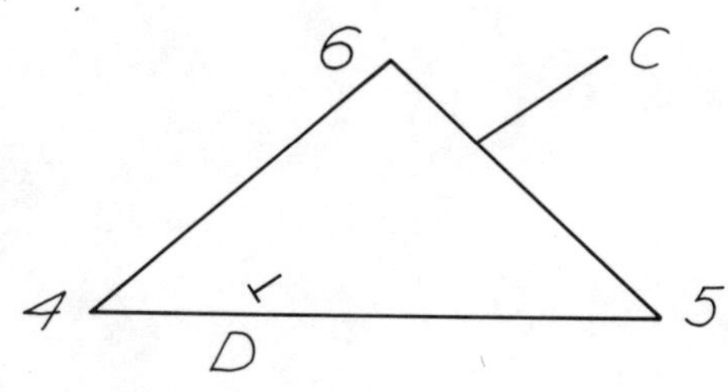

H

F

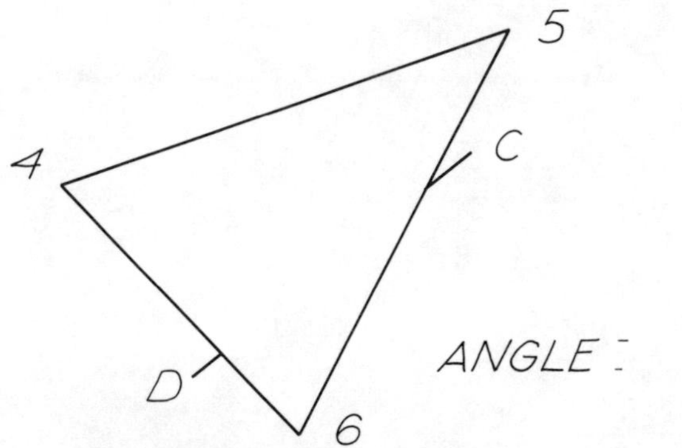

JHE

1 COMPLETE THE FRONT VIEW OF THE CONE CONTAINING LINE AB.

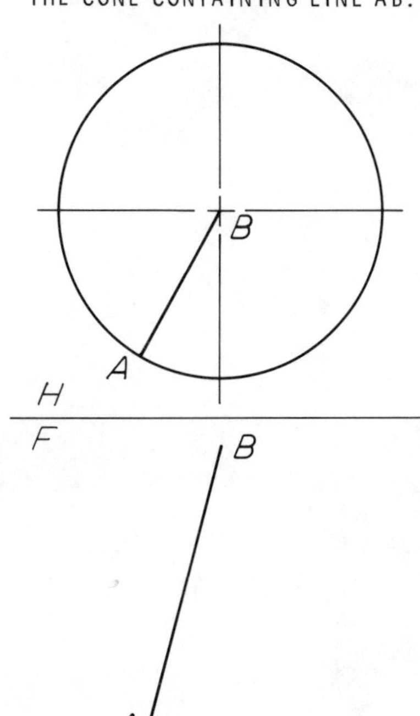

2 FIND THE TRUE LENGTH OF LINE CD IN THE FRONT VIEW BY REVOLVING POINT D ABOUT AN AXIS THROUGH POINT C.

C
D
H
F
SLOPE =
C
D

3 FIND THE TRUE LENGTH OF LINE EG IN THE TOP VIEW BY REVOLVING POINT G ABOUT AN AXIS THROUGH POINT E.

ANGLE WITH F :

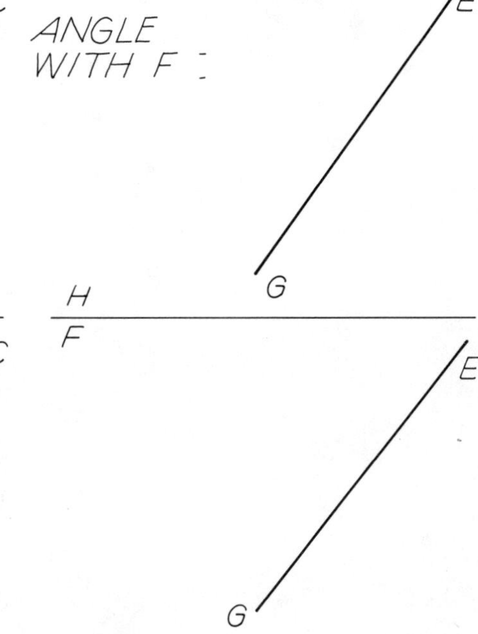

4 FIND THE TRUE SIZE OF PLANE 1-2-3 BY REVOLUTION. HINT: FIND EDGE VIEW OF PLANE BY AUXILIARY OFF THE TOP VIEW, THEN ROTATE.

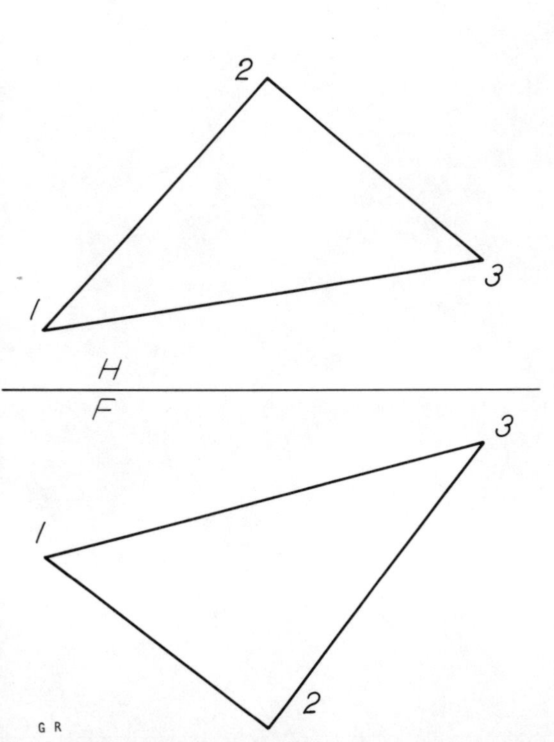

5 BY DOUBLE REVOLUTION, FIND THE TRUE SIZE VIEW OF PLANE ABC IN THE TOP VIEW. REVOLVE THE TOP VIEW FIRST.

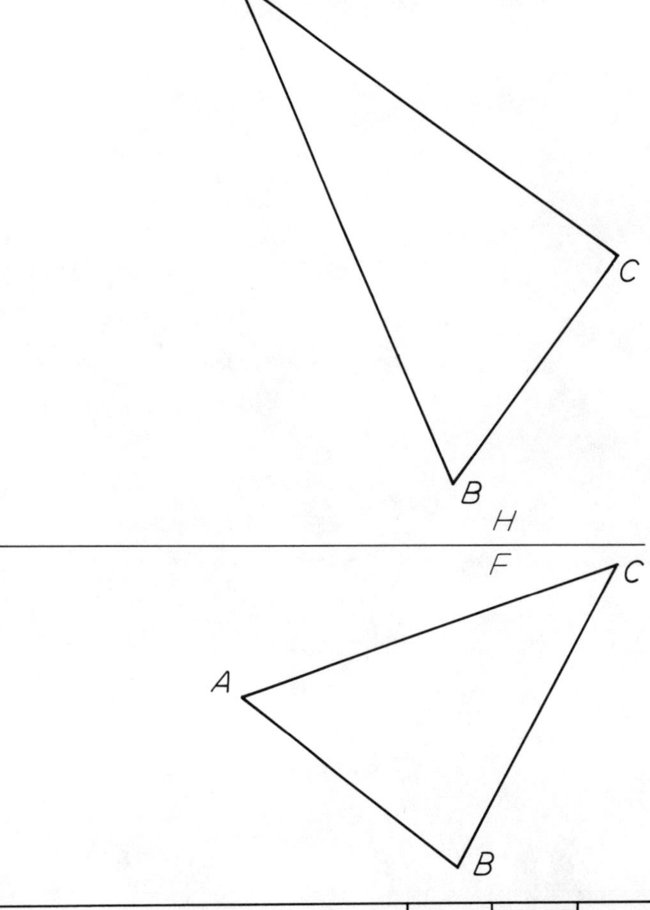

G R

FIND THE ANGLES BETWEEN THE INTERSECTING PLANES IN PROBLEMS 1, 2, & 3 BY REVOLUTION. PASS THE PLANE TO BE REVOLVED THROUGH POINT A IN EACH PROBLEM.

1 ANGLE =

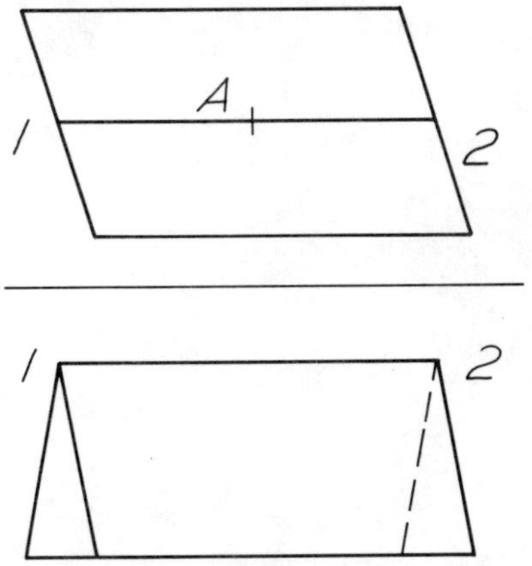

2 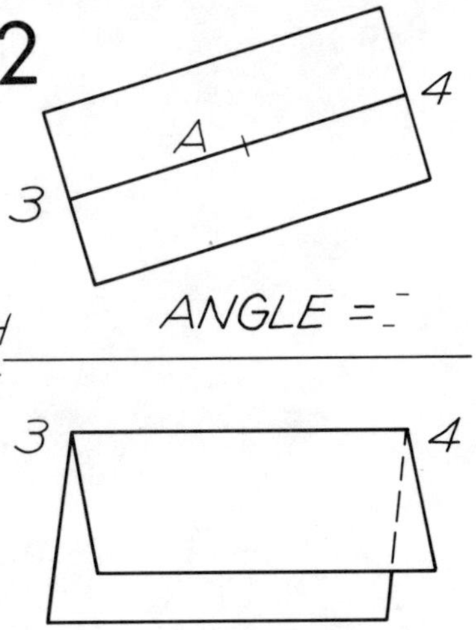 ANGLE =

3 ANGLE =

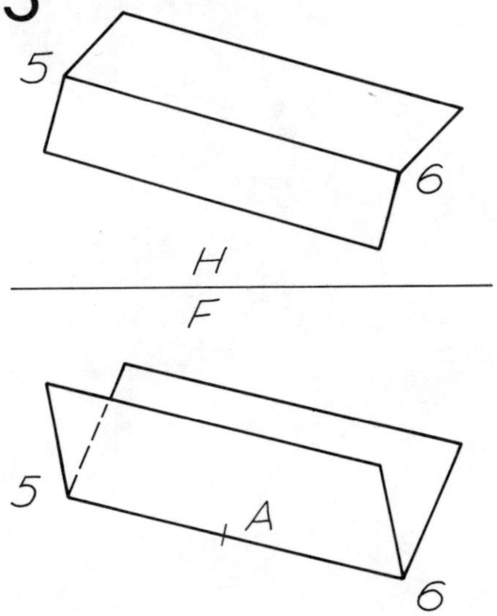

4 AN ENGINE MOUNT FOR A HELICOPTER IS SHOWN BELOW. (NOTE THAT SOME MEMBERS HAVE BEEN OMITTED AND THE JOINTS HAVE BEEN SIMPLIFIED FOR CLARITY.)

A. FIND THE ANGLE BETWEEN PLANES 1-2-4 AND 1-2-3 BY REVOLUTION. PROJECT AN AUXILIARY VIEW FROM THE TOP VIEW.

B. FIND THE ANGLE BETWEEN PLANES 1-3-5 AND 1-5-4. PASS THE PLANE TO BE REVOLVED THROUGH POINT A AND FIND THE TRUE ANGLE IN THE FRONT VIEW.

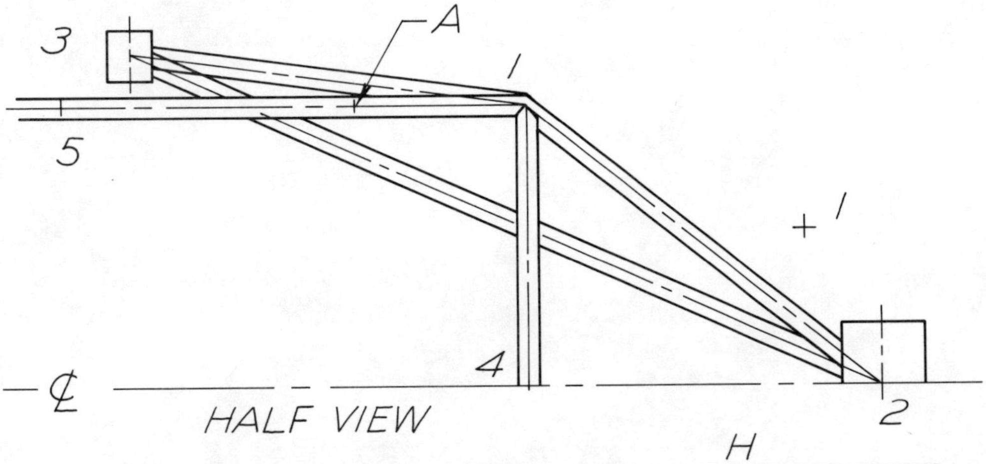

HALF VIEW

ANGLES

PROB A

PROB B

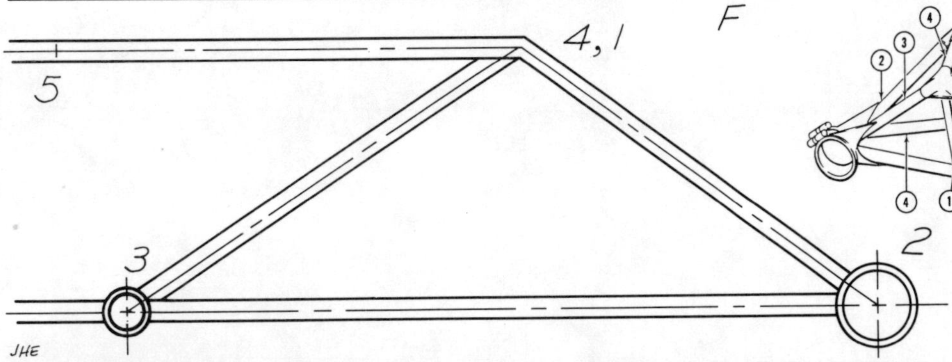

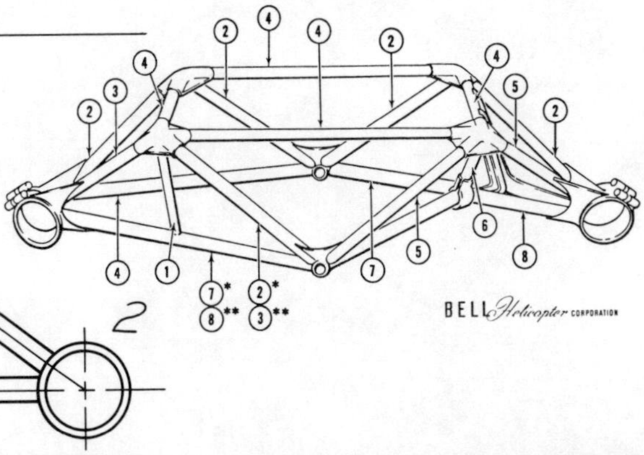

BELL Helicopter CORPORATION

JHE

Graphics & Geometry

NAME

FILE SEC DATE

MIN. GRADE

REVOLUTION

1 ROTATE POINT O ABOUT LINE 1-2 INTO ITS HIGHEST POSITION. SHOW POINT O IN ALL VIEWS.

O
1
2
H
F
O
1
2

2 ROTATE POINT O ABOUT LINE 3-4 INTO ITS MOST FORWARD POSITION. SHOW POINT O IN ALL VIEWS.

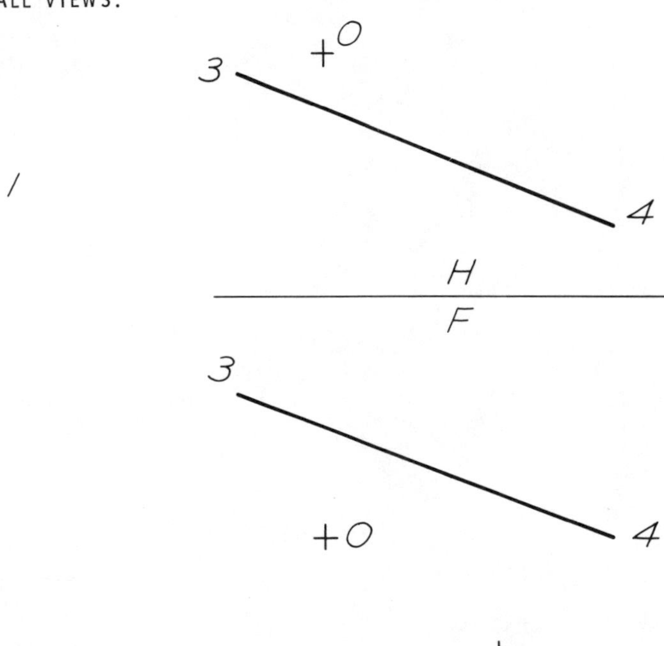

2
1

3

THE SPACE CAPSULE IS EQUIPPED WITH A CAMERA (POINT P) THAT MUST BE ROTATED (ROLLED) ABOUT THE AXIS OF THE CAPSULE (LINE 5-6) TO ITS LOWEST POSITION IN ORDER TO PHOTOGRAPH THE LUNAR SURFACE.

ROTATE THIS POINT TO ITS LOWEST POSITION AND SHOW THE POINT IN ALL VIEWS.

WHAT IS THE ANGLE OF ROTATION REQUIRED FOR THIS MANUEVER?

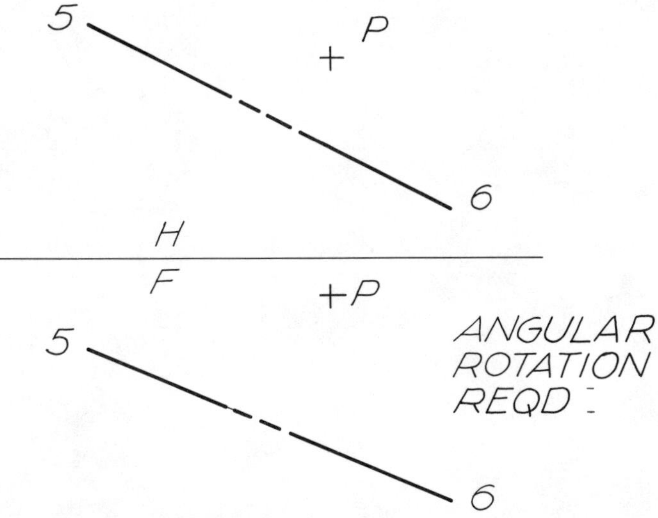

2

ANGULAR ROTATION REQD

courtesy of Bell Aerosystems Company

JHE

INTERSECTIONS

1 FIND THE INTERSECTIONS BETWEEN THESE SHEET METAL PARTS

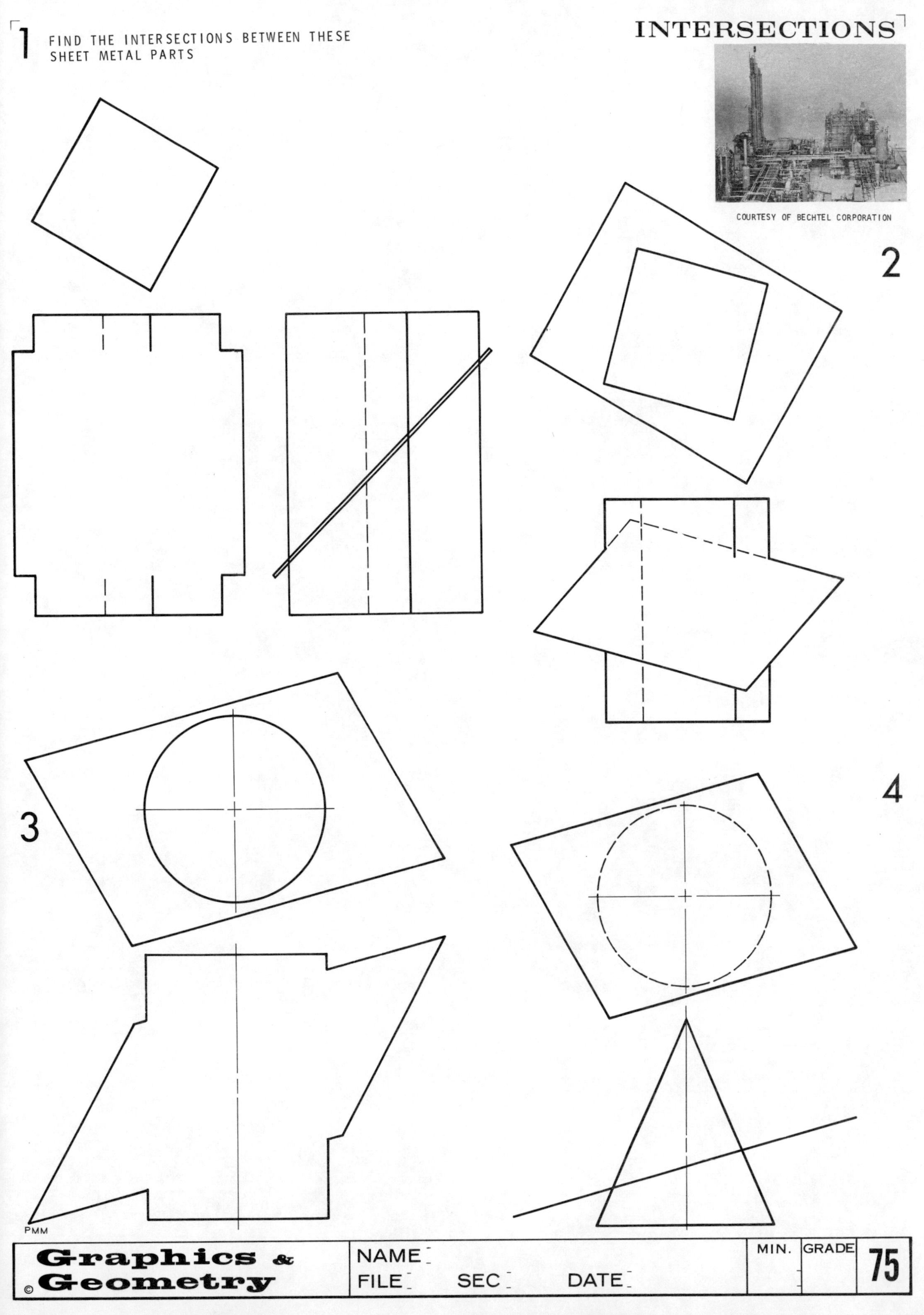

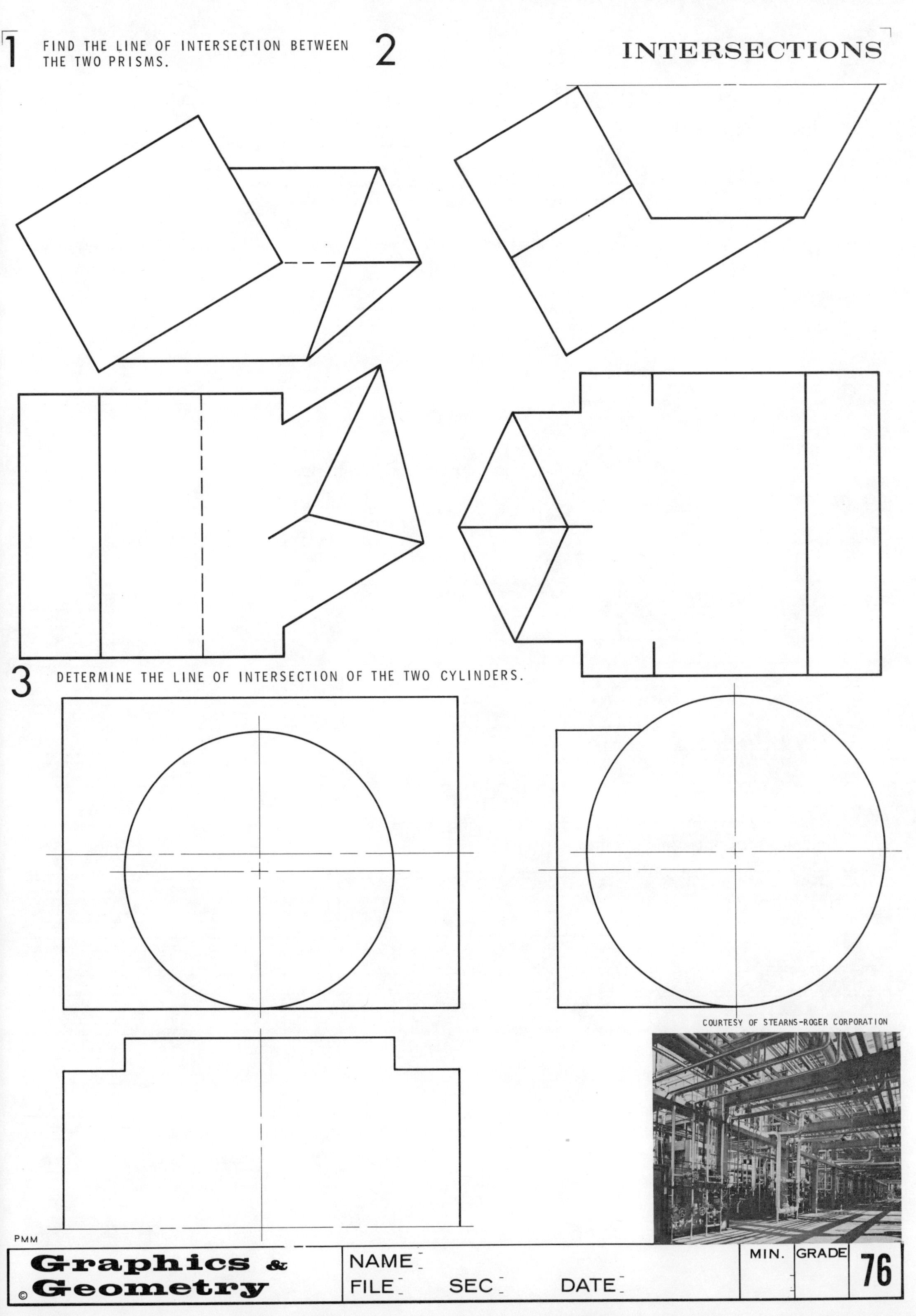
1 FIND THE LINE OF INTERSECTION BETWEEN THE TWO PRISMS.
2
3 DETERMINE THE LINE OF INTERSECTION OF THE TWO CYLINDERS.
COURTESY OF STEARNS-ROGER CORPORATION
PMM
Graphics & Geometry
NAME
FILE
SEC
DATE
MIN.
GRADE

LETTERING GUIDELINES

These guidelines can be used to underlay tracing paper or other transluscent papers when lettering a drawing. Vertical as well as horizontal guidelines are given.

One-eighth inch guidelines

MILD STEEL

Tolerances

2.177
2.174

Notes—Half-spacing between lines of lettering.

ALL FILLETS &
ROUNDS .25 R

Guidelines for upper and lower-case letters.

Caps and lower-case

Guidelines for common fractions;

$2\frac{1}{2}$, 13'-$7\frac{1}{4}$

1 FIND THE LINE OF INTERSECTION BETWEEN THE TWO CYLINDERS.

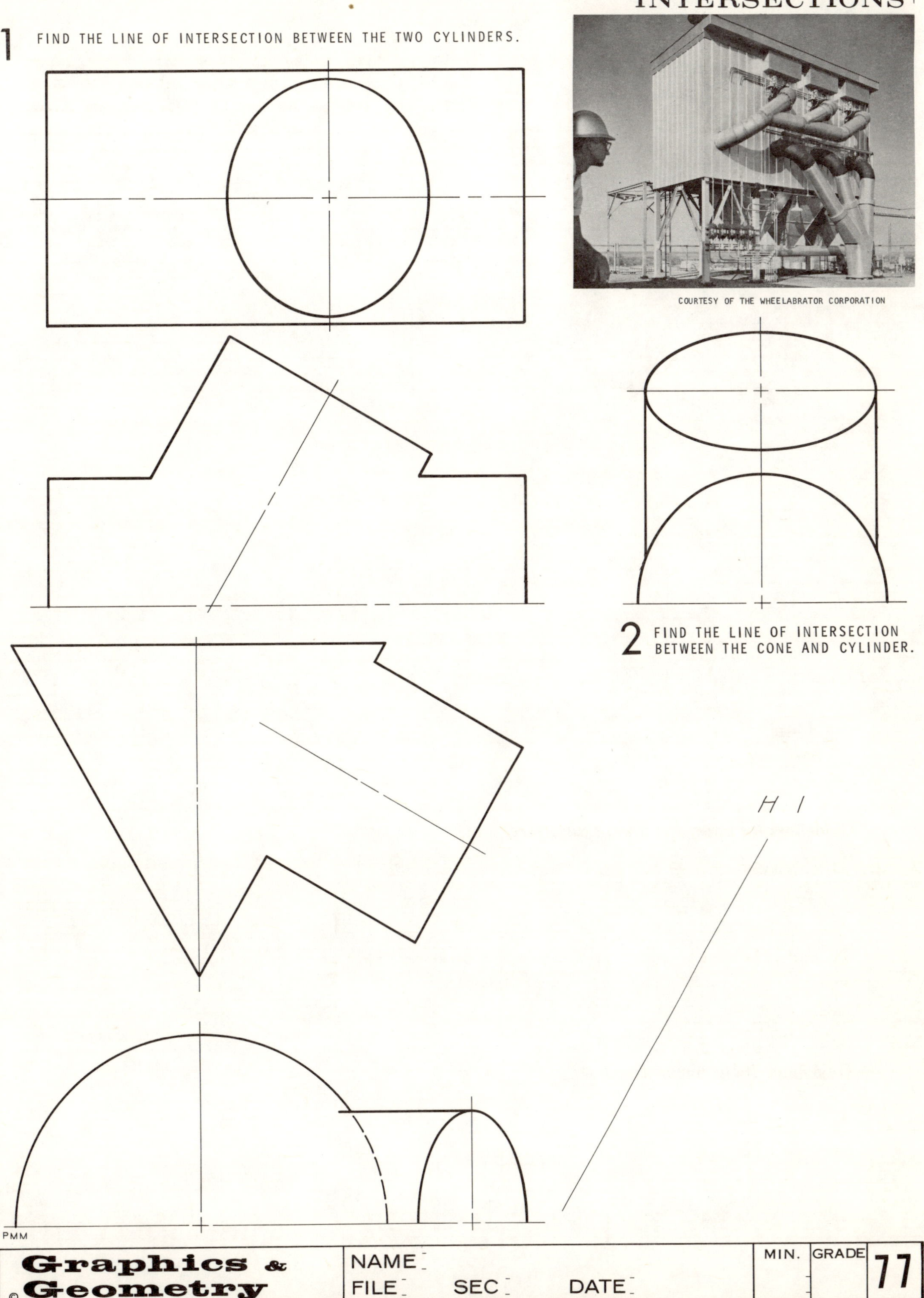

COURTESY OF THE WHEELABRATOR CORPORATION

2 FIND THE LINE OF INTERSECTION BETWEEN THE CONE AND CYLINDER.

PMM

LETTERING GUIDELINES

These guidelines can be used to underlay tracing paper or other transluscent papers when lettering a drawing. Vertical as well as horizontal guidelines are given.

One-eighth inch guidelines

MILD STEEL

Tolerances

2.177
2.174

Notes—Half-spacing between lines of lettering.

ALL FILLETS &
ROUNDS .25 R

Guidelines for upper and lower-case letters.

Caps and lower-case

Guidelines for common fractions;

$2\frac{1}{2}$, $13' - 7\frac{1}{4}$

1 LAY OUT AN INSIDE FLAT PATTERN OF THE RECTANGULAR DUCT. BEGIN THE PATTERN ON LINE 0-1. NUMBER THE POINTS.

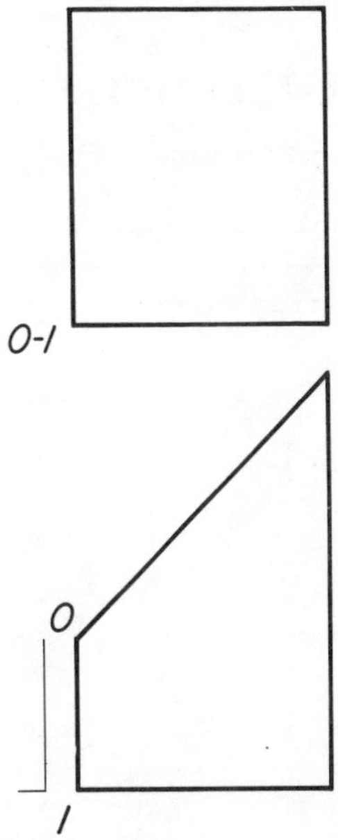

0-1

0

1

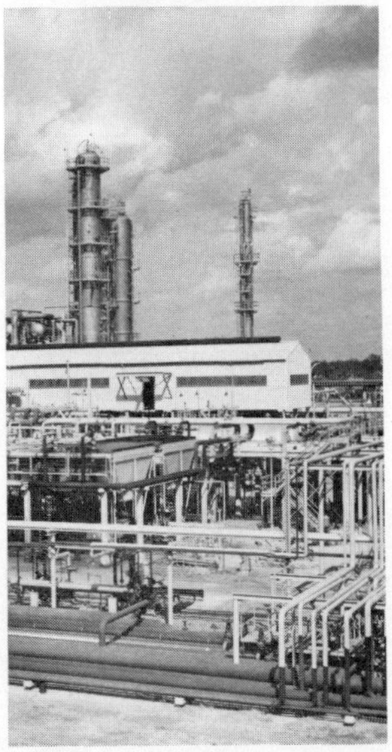

COURTESY OF PHILLIPS PETROLEUM CO.

2 A SQUARE HOPPER IS LOADED FROM A CONICAL FEEDER. LAY OUT A HALF, S. INSIDE PATTERN OF THE HOPPER. BEGIN THE PATTERN ON LINE OP. NUMBER THE OP. POINTS.

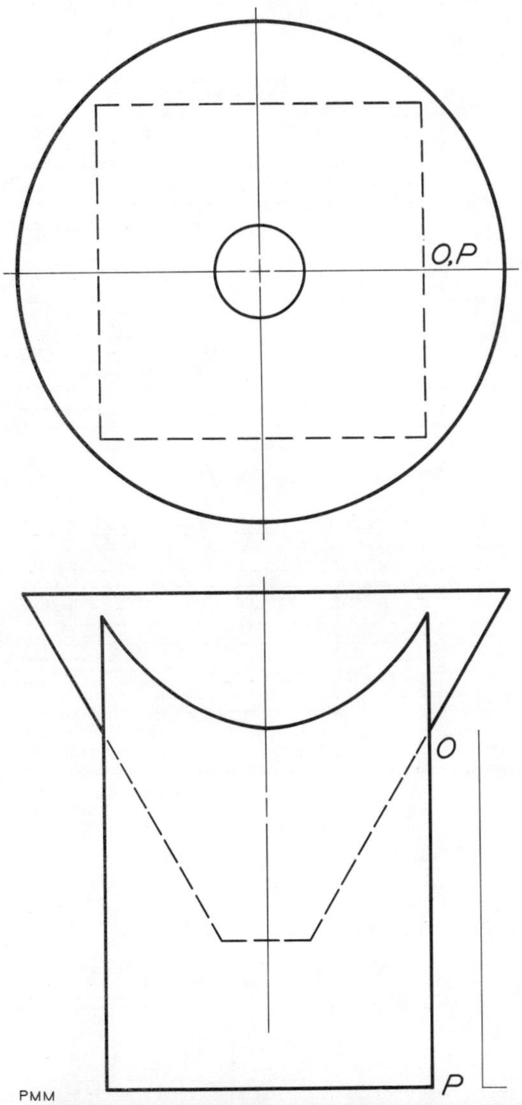

3. DEVELOP A HALF INSIDE PATTERN OF THE CYLINDRICAL DUCT. INCLUDE POINTS 1 THROUGH 7 ON THE PATTERN.

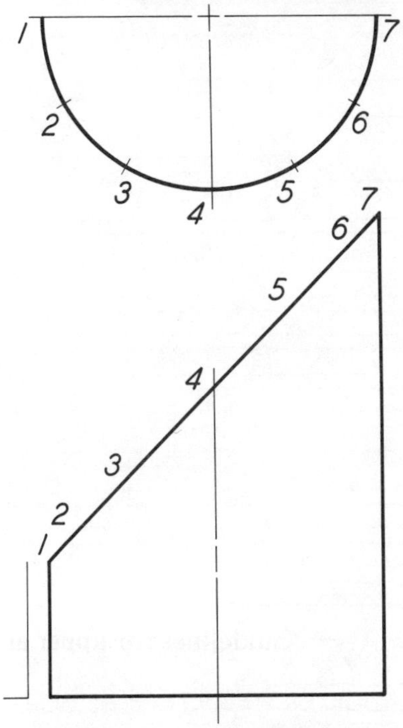

1 2 3 4 5 6 7

7 6 5 4 3 2 1

O,P

O

P

COURTESY OF SYNTRON DIVISION OF FMC CORPORATION

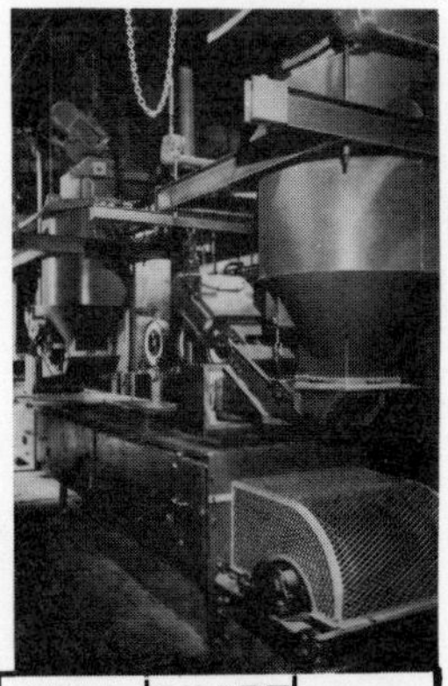

PMM

LETTERING GUIDELINES

These guidelines can be used to underlay tracing paper or other transluscent papers when lettering a drawing. Vertical as well as horizontal guidelines are given.

One-eighth inch guidelines

MILD STEEL

Tolerances

2.177
2.174

Notes—Half-spacing between lines of lettering.

ALL FILLETS &
ROUNDS .25 R

Guidelines for upper and lower-case letters.

Caps and lower-case

Guidelines for common fractions;

$2\frac{1}{2}$, $13' - 7\frac{1}{4}$

DEVELOPMENTS

1 DETERMINE THE RIGHT SECTION AND DEVELOP THE INSIDE FLAT PATTERN OF THE INCLINED CIRCULAR CYLINDER. BEGIN THE PATTERN ON LINE AA'. LETTER THE POINTS.

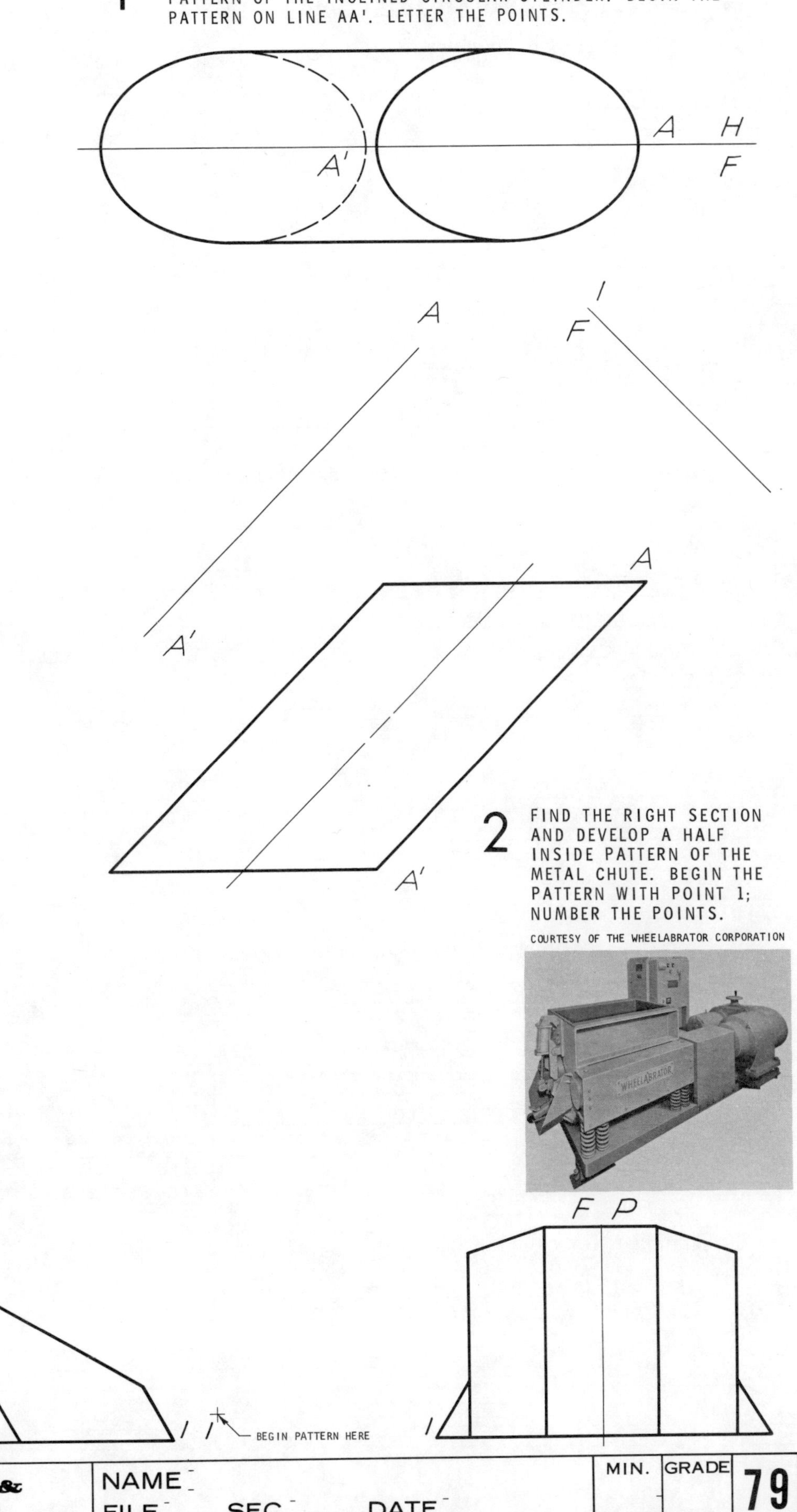

2 FIND THE RIGHT SECTION AND DEVELOP A HALF INSIDE PATTERN OF THE METAL CHUTE. BEGIN THE PATTERN WITH POINT 1; NUMBER THE POINTS.

COURTESY OF THE WHEELABRATOR CORPORATION

PMM

LETTERING GUIDELINES

These guidelines can be used to underlay tracing paper or other transluscent papers when lettering a drawing. Vertical as well as horizontal guidelines are given.

One-eighth inch guidelines

MILD STEEL

Tolerances

2.177
2.174

Notes—Half-spacing between lines of lettering.

ALL FILLETS &
ROUNDS .25 R

Guidelines for upper and lower-case letters.

Caps and lower-case

Guidelines for common fractions;

$2\frac{1}{2}$, 13'-$7\frac{1}{4}$

1 LAY OUT AN INSIDE, HALF PATTERN OF THE CONE. START THE PATTERN WITH LINE 0-1. NUMBER THE POINTS.

2 DEVELOP AN INSIDE, HALF PATTERN OF THE HOPPER OF A PORTO-SPREADER. START THE PATTERN ON LINE 0-2. NUMBER THE POINTS.

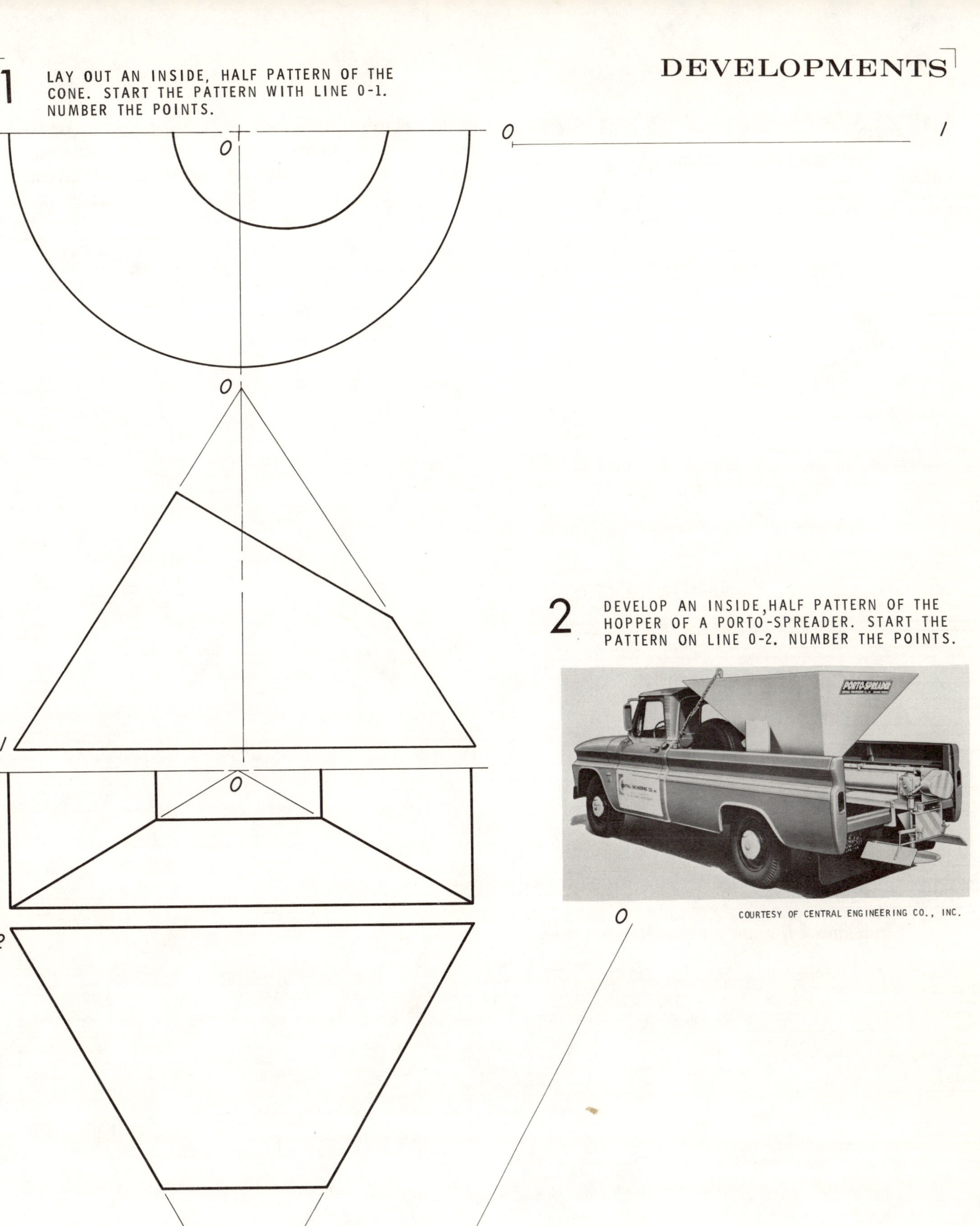

COURTESY OF CENTRAL ENGINEERING CO., INC.

LETTERING GUIDELINES

These guidelines can be used to underlay tracing paper or other transluscent papers when lettering a drawing. Vertical as well as horizontal guidelines are given.

One-eighth inch guidelines

MILD STEEL

Tolerances

2.177
2.174

Notes—Half-spacing between lines of lettering.

ALL FILLETS &
ROUNDS .25 R

Guidelines for upper and lower-case letters.

Caps and lower-case

Guidelines for common fractions;

$2\frac{1}{2}$, 13'-$7\frac{1}{4}$

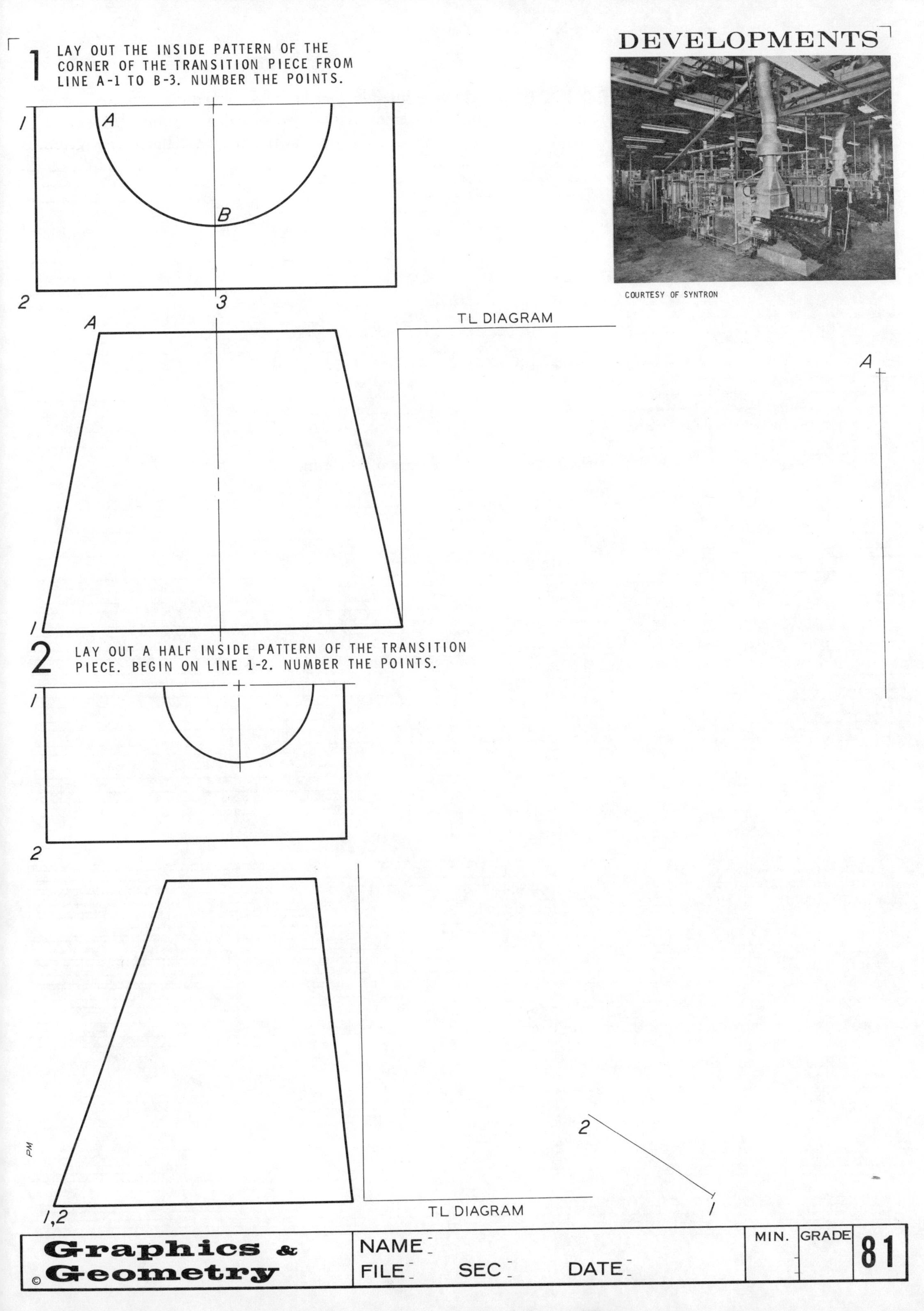
DEVELOPMENTS
1 LAY OUT THE INSIDE PATTERN OF THE CORNER OF THE TRANSITION PIECE FROM LINE A-1 TO B-3. NUMBER THE POINTS.
A
B
1
2
3
A
1
TL DIAGRAM
A
COURTESY OF SYNTRON
2 LAY OUT A HALF INSIDE PATTERN OF THE TRANSITION PIECE. BEGIN ON LINE 1-2. NUMBER THE POINTS.
1
2
1,2
TL DIAGRAM
2
1
Graphics & Geometry
©
NAME
FILE
SEC
DATE
MIN.
GRADE
81

LETTERING GUIDELINES

These guidelines can be used to underlay tracing paper or other transluscent papers when lettering a drawing. Vertical as well as horizontal guidelines are given.

One-eighth inch guidelines

MILD STEEL

Tolerances

2.177
2.174

Notes—Half-spacing between lines of lettering.

ALL FILLETS &
ROUNDS .25 R

Guidelines for upper and lower-case letters.

Caps and lower-case

Guidelines for common fractions;

$2\frac{1}{2}$, $13' = 7\frac{1}{4}$

CONCURRENT, COPLANAR VECTORS

1 NEWTON'S SECOND LAW STATES THAT THE ACCELERATION OF AN OBJECT IS PROPORTIONAL TO AND IN THE DIRECTION OF THE RESULTANT FORCE. FIND THE RESULTANT FORCE OF THE VECTOR SYSTEM SHOWN. START WITH VECTOR A AND WORK ALPHABETICALLY.

A. SOLVE BY THE PARALLELOGRAM METHOD.

B. SOLVE BY THE POLYGON METHOD.

SCALE: 1=400 LBS

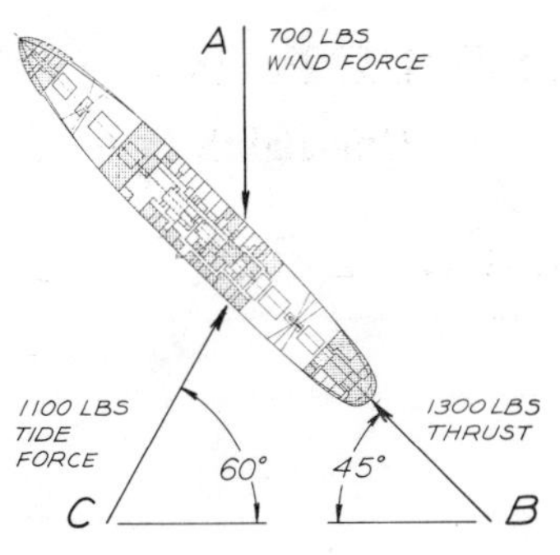

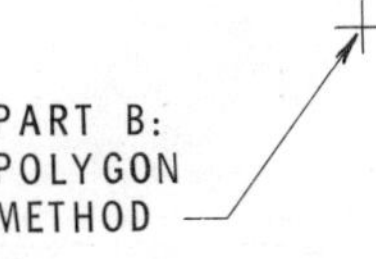

ANS =

ANS =

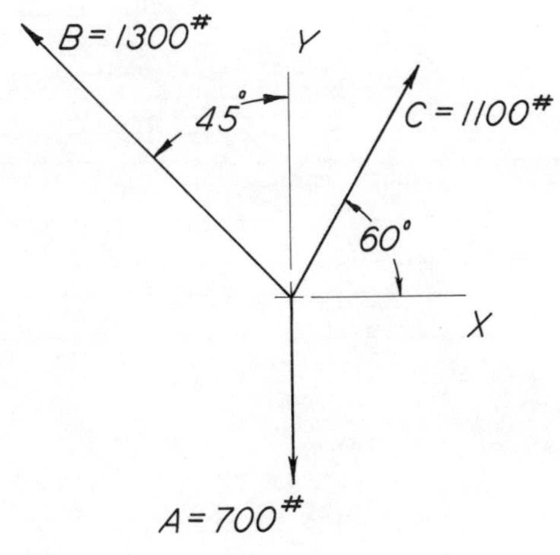

$\Sigma F_X = -B \cos 45° + C \cos 60° = R_X$

$=$

$R_X =$

$\Sigma F_Y = B \sin 45° + C \sin 60° - A = R_Y$

$=$

$R_Y =$

$\theta = \arctan \frac{R_Y}{R_X} =$

$R = \sqrt{R_X^2 + R_Y^2} =$

$R =$

2 SOLVE THE ABOVE PROBLEM MATHEMATICALLY. COMPARE THE ANSWERS WITH THOSE OBTAINED GRAPHICALLY.

JTC

Graphics & Geometry ©	NAME FILE SEC DATE	MIN.	GRADE	82

LETTERING GUIDELINES

These guidelines can be used to underlay tracing paper or other transluscent papers when lettering a drawing. Vertical as well as horizontal guidelines are given.

One-eighth inch guidelines

MILD STEEL

Tolerances

2.177
2.174

Notes—Half-spacing between lines of lettering.

ALL FILLETS &
ROUNDS .25 R

Guidelines for upper and lower-case letters.

Caps and lower-case

Guidelines for common fractions;

$2\frac{1}{2}$, 13' = $7\frac{1}{4}$

VECTOR ANALYSIS

1 A WEIGHT IS SUPPORTED FROM THE WALL AS SHOWN. DETERMINE LOADS CARRIED IN EACH OF THE STRUCTURAL MEMBERS.

SCALE: 1 = 400#

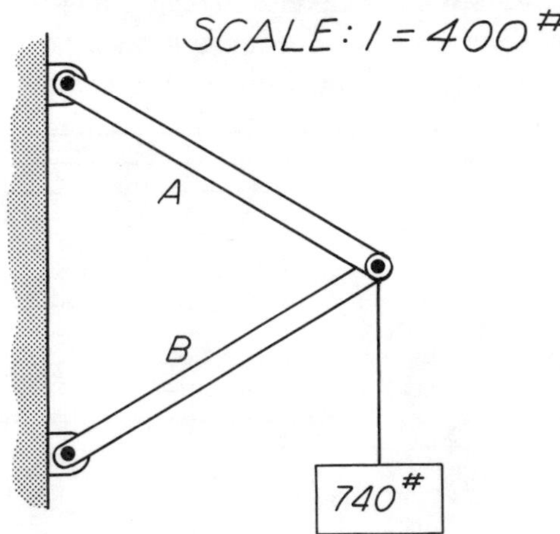

BEGIN 740#
VECTOR HERE

COURTESY OF PACIFIC HOIST

MEMBER	LOAD	TENSION OR COMPRESSION
A		
B		

BEGIN 1000#
VECTOR HERE

2 IF A CABLE IS STRETCHED OVER A "FRICTIONLESS" PULLEY, THE TENSION IN THE CABLE IS THE SAME ON BOTH SIDES. ANALYZE THE FORCES ACTING ON THE END OF THE BOOM. START WITH THE VERTICAL LOAD AND PROCEED IN A COUNTER CLOCKWISE MANNER.

SCALE: 1 = 200#

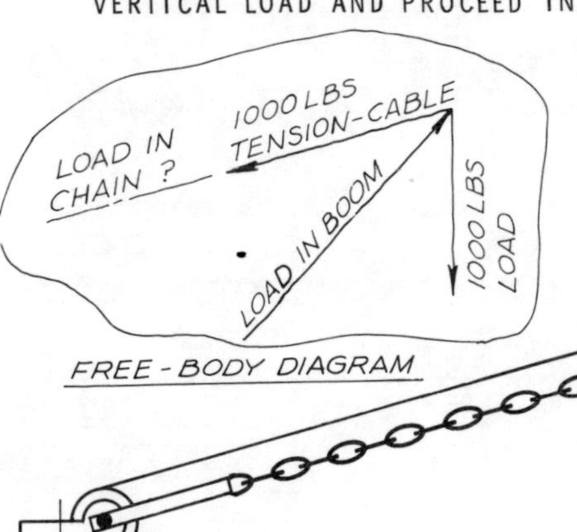

FREE-BODY DIAGRAM

1000 LBS

MEMBER	LOAD	TENSION OR COMPRESSION
CHAIN		
BOOM		

LETTERING GUIDELINES

These guidelines can be used to underlay tracing paper or other transluscent papers when lettering a drawing. Vertical as well as horizontal guidelines are given.

One-eighth inch guidelines

MILD STEEL

Tolerances

2.177
2.174

Notes—Half-spacing between lines of lettering.

ALL FILLETS &
ROUNDS .25 R

Guidelines for upper and lower-case letters.

Caps and lower-case

Guidelines for common fractions;

$2\frac{1}{2}$, 13' = $7\frac{1}{4}$

BEAM ANALYSIS

SUPPORT AND MEMBER SIZES MAY BE DETERMINED DURING THE ANALYSIS STAGE OF THE DESIGN PROCESS BY USING BOW'S NOTATION.

1 SOLVE GRAPHICALLY FOR THE REACTIONS ON THE SWING SET.

2 LOCATE GRAPHICALLY THE POINT OF APPLICATION OF THE EQUILIBRANT FORCE AND INDICATE ITS DISTANCE FROM THE CENTER OF THE LEFT COLUMN.

3 SOLVE THE ENTIRE PROBLEM MATHEMATICALLY AS A CHECK. USE THE SPACE PROVIDED.

PHOTO BY JTC

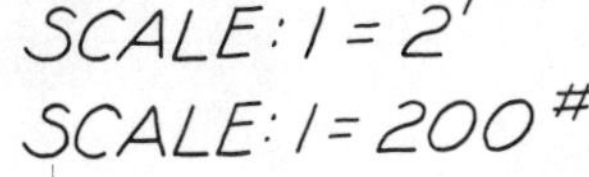

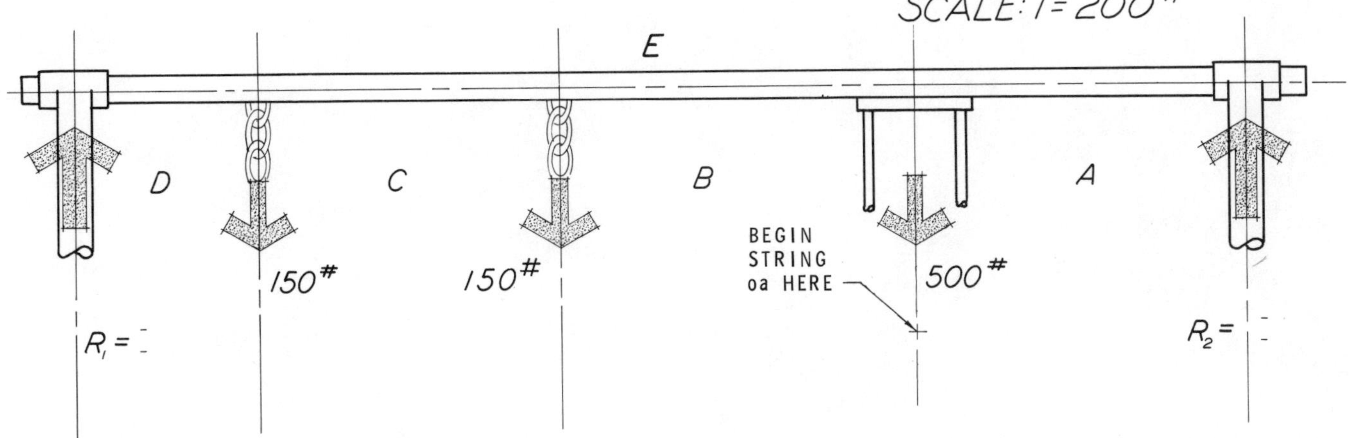

STRING DIAGRAM POLE POINT

O

MATHEMATICAL SOLUTION

$\Sigma M_{R_1} = 0 =$

$R_2 =$

$\Sigma M_{R_2} = 0 =$

$R_1 =$

CHECK:

$R_1 + R_2 =$

JTC

Graphics & Geometry ©	NAME FILE SEC DATE	MIN.	GRADE	84

Graphics & Geometry ©	NAME:	MIN.	GRADE	
	FILE: SEC: DATE:			

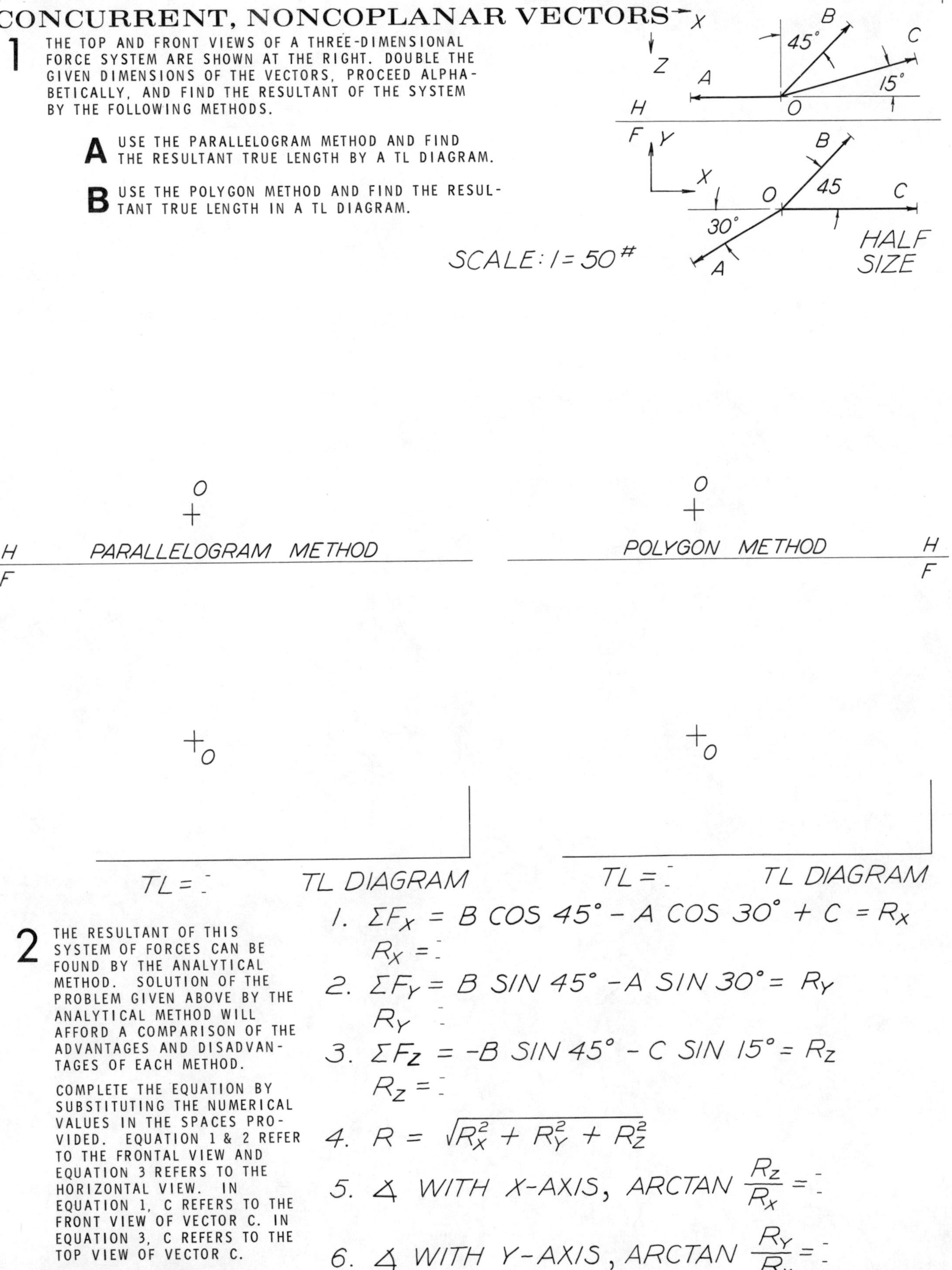

CONCURRENT, NONCOPLANAR VECTORS

1 THE TOP AND FRONT VIEWS OF A THREE-DIMENSIONAL FORCE SYSTEM ARE SHOWN AT THE RIGHT. DOUBLE THE GIVEN DIMENSIONS OF THE VECTORS, PROCEED ALPHABETICALLY, AND FIND THE RESULTANT OF THE SYSTEM BY THE FOLLOWING METHODS.

A USE THE PARALLELOGRAM METHOD AND FIND THE RESULTANT TRUE LENGTH BY A TL DIAGRAM.

B USE THE POLYGON METHOD AND FIND THE RESULTANT TRUE LENGTH IN A TL DIAGRAM.

2 THE RESULTANT OF THIS SYSTEM OF FORCES CAN BE FOUND BY THE ANALYTICAL METHOD. SOLUTION OF THE PROBLEM GIVEN ABOVE BY THE ANALYTICAL METHOD WILL AFFORD A COMPARISON OF THE ADVANTAGES AND DISADVANTAGES OF EACH METHOD.

COMPLETE THE EQUATION BY SUBSTITUTING THE NUMERICAL VALUES IN THE SPACES PROVIDED. EQUATION 1 & 2 REFER TO THE FRONTAL VIEW AND EQUATION 3 REFERS TO THE HORIZONTAL VIEW. IN EQUATION 1, C REFERS TO THE FRONT VIEW OF VECTOR C. IN EQUATION 3, C REFERS TO THE TOP VIEW OF VECTOR C.

1. $\Sigma F_X = B \cos 45° - A \cos 30° + C = R_X$

   $R_X =$

2. $\Sigma F_Y = B \sin 45° - A \sin 30° = R_Y$

   R_Y

3. $\Sigma F_Z = -B \sin 45° - C \sin 15° = R_Z$

   $R_Z =$

4. $R = \sqrt{R_X^2 + R_Y^2 + R_Z^2}$

5. $\measuredangle$ WITH X-AXIS, ARCTAN $\frac{R_Z}{R_X} =$

6. $\measuredangle$ WITH Y-AXIS, ARCTAN $\frac{R_Y}{R_X} =$

Graphics & Geometry ©	NAME:	MIN.	GRADE	
	FILE: SEC: DATE:			

VECTOR ANALYSIS - SPECIAL CASE

SHOWN IN THE PHOTOGRAPH IS A CRANE WHICH HAS BEEN USED TO RAISE AND LOWER MINIATURE SUBMARINES OVER THE SIDE OF A SHIP.

A. DETERMINE THE LOADS IN THE HYDRAULIC CYLINDERS AND IN THE CABLE GRAPHICALLY.

B. SOLVE MATHEMATICALLY ON A SEPARATE SHEET.

COURTESY OF NORTHER LINE MACHINE & ENGINEERING COMPANY

BEGIN FRONT VIEW OF LOAD HERE

SCALE: 1=1000 LBS

MEMBER	LOAD	TENSION OR COMPRESSION
A		
B		
C		

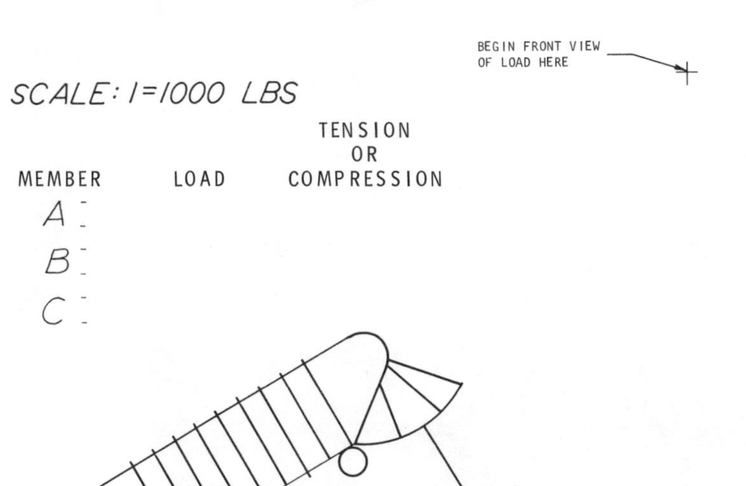

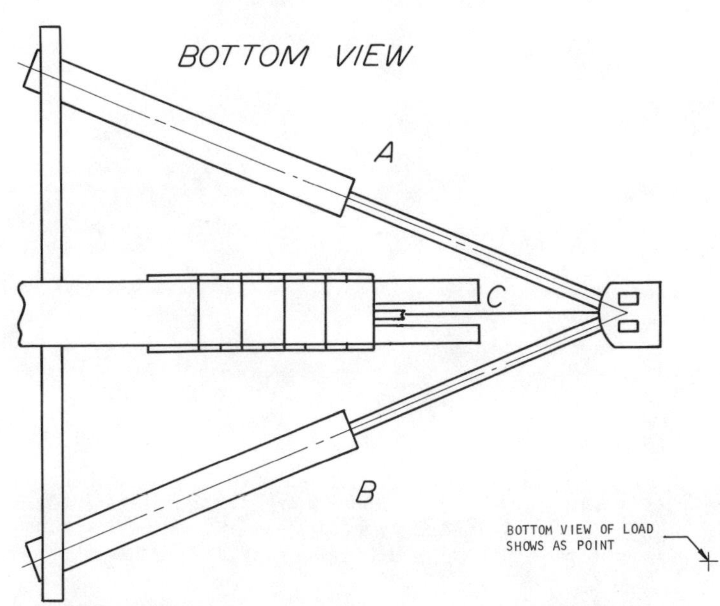

BOTTOM VIEW OF LOAD SHOWS AS POINT

JTC

Graphics & ©Geometry	NAME FILE SEC DATE	MIN.	GRADE	

NONCOPLANAR VECTORS

DETERMINE THE LOADS IN THE CABLES THAT SUPPORT THE 160 LB TRAFFIC SIGNAL. (HINT: RESOLVE CABLES A AND C INTO ONE PLANE.)

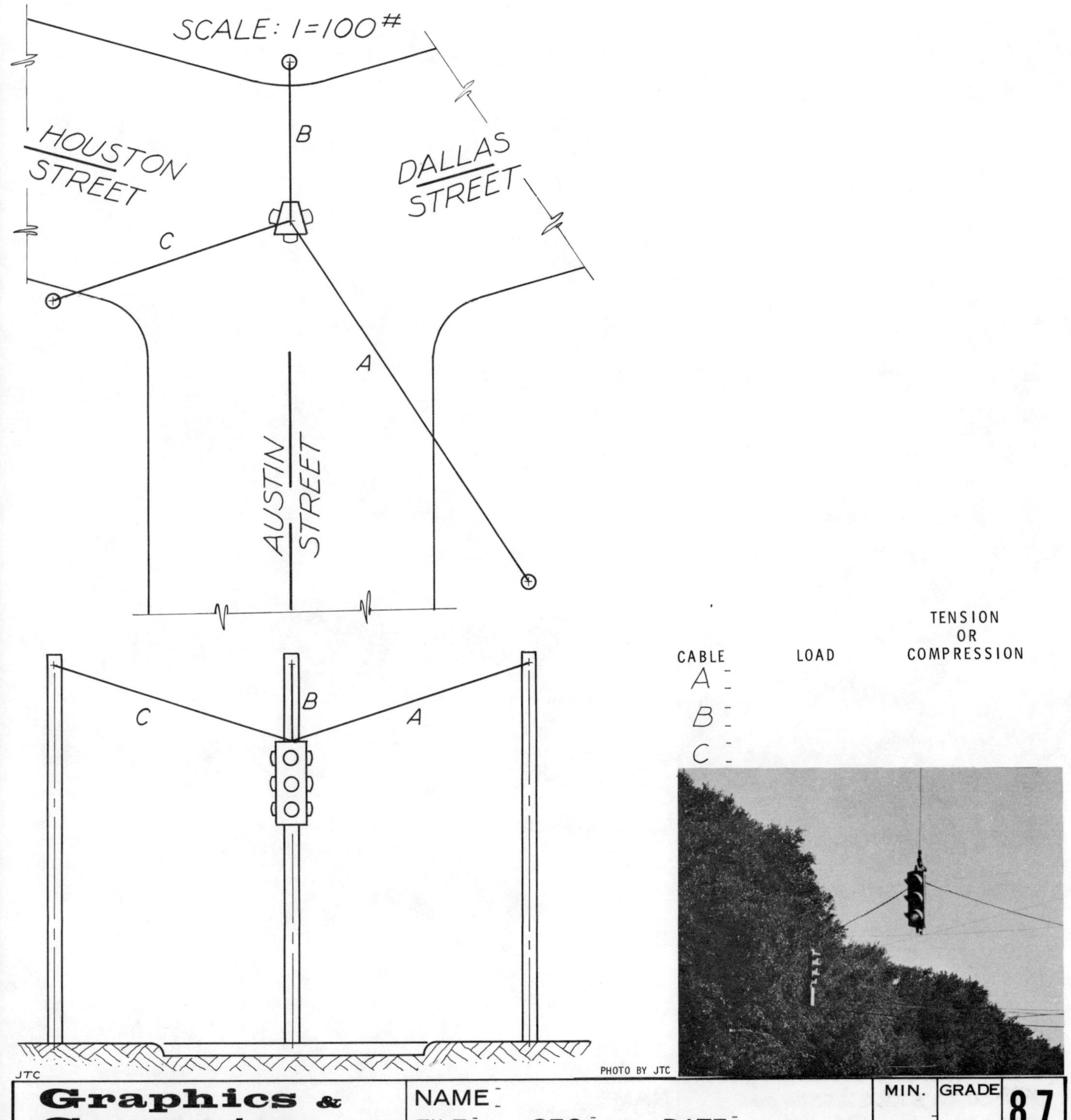

CABLE	LOAD	TENSION OR COMPRESSION
A		
B		
C		

TRUSS ANALYSIS

TO PROPERLY SIZE THE MEMBERS OF A TRUSS, THE LOADS THAT EACH WILL CARRY MUST BE DETERMINED. ANALYZE THE SAWTOOTH TRUSS SHOWN.

1 BEGINNING WITH JOINT 1, DRAW THE VECTOR POLYGON FOR EACH JOINT.

2 CONSTRUCT A MAXWELL DIAGRAM. NOTICE THAT THIS DIAGRAM IS A SUPERPOSITION OF THE VECTOR POLYGONS OF EACH JOINT.

3 ON A SEPARATE SHEET, SOLVE MATHEMATICALLY.

SCALE: 1 = 4000#

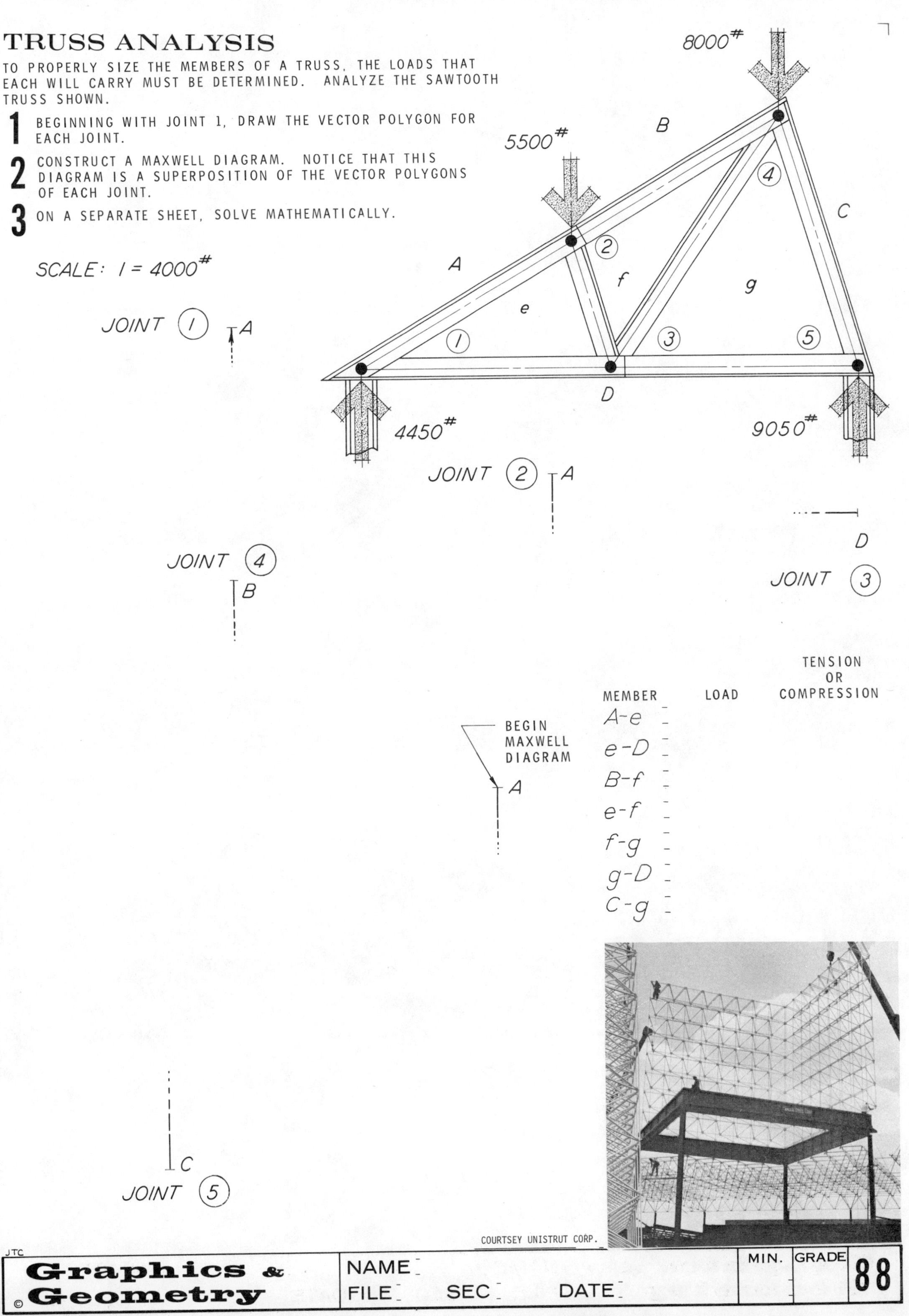

MEMBER	LOAD	TENSION OR COMPRESSION
A-e		
e-D		
B-f		
e-f		
f-g		
g-D		
C-g		

COURTSEY UNISTRUT CORP.

BAR GRAPHS

DURING THE CURRENT YEAR SOME PLATING METAL PRICES HAVE INCREASED RAPIDLY. THE PRICE INCREASES ARE SHOWN IN THE TABLE BELOW AND IN THE FORM OF A BAR GRAPH AS SKETCHED TO THE RIGHT. WITH INSTRUMENTS DRAW AND ENLARGE THE BAR GRAPH IN THE SPACE BELOW. GRAPHS DRAWN FOR WRITTEN REPORTS SHOULD BE IN 2:3 PROPORTIONS SO THAT 35MM SLIDES CAN BE MADE FROM THEM FOR ORAL PRESENTATIONS.

METAL	PRICE INCREASE (PER CENT)
NICKEL	22
LEAD	58
TIN	70
SILVER	97

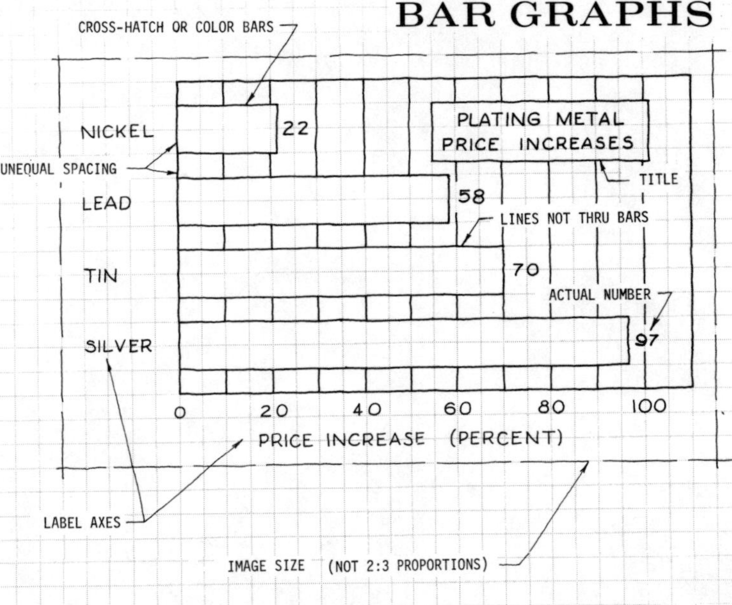

3

IMAGE SIZE

2

1.5

IMAGE SIZE

1

53MM ≈ 3

35MM ≈ 2

35MM SLIDE

A 35MM CAMERA CAN BE FOCUSED SO THAT ANY 2:3 OR 3:2 IMAGE WILL FILL THE VIEWFINDER AS ILLUSTRATED IN THE FIGURE TO THE LEFT. WHEN DRAWING THE GRAPH IN THE SPACE BELOW ALL INFORMATION SHOULD BE INSIDE THE IMAGE LINES. THE INFORMATION SHOULD FILL THE SPACE BUT SHOULD NOT BE CROWDED.

CONSTRUCT THE 2:3 IMAGE SIZE USING THE STARTING POINT GIVEN BELOW.

BEGIN HERE

JTD

Graphics & Geometry	NAME	MIN.	GRADE
©	FILE SEC DATE		

BROKEN-LINE GRAPHS

BROKEN-LINE GRAPH

A DEALER WHO SELLS SPORTS AND RECREATIONAL VEHICLES HAS RECORDED HIS SALES INFORMATION FOR BICYCLES AND GO-KARTS SINCE 1956. THE SALES INFORMATION FOR BOTH VEHICLES IS GIVEN IN THE TABLE AND SKETCHED BROKEN-LINE GRAPH SHOWN BELOW.

VEHICLE SALES

YEAR	BICYCLES	GO-CARTS
1956	50	10
1960	70	90
1964	60	140
1968	80	30
1972	120	90
1976	180	130

USING INSTRUMENTS DRAW THE GRAPH IN 3:2 PROPORTIONS SO THAT THE IMAGE WILL FILL THE SPACE TO THE LEFT. THE SKETCHED GRAPH IS NOT TO SCALE BUT DOES CONTAIN THE ELEMENTS NECESSARY FOR A GOOD GRAPH. CONNECT THE DATA POINTS WITH STRAIGHT LINE SEGMENTS WITHOUT PENETRATING THE POINT SYMBOL.

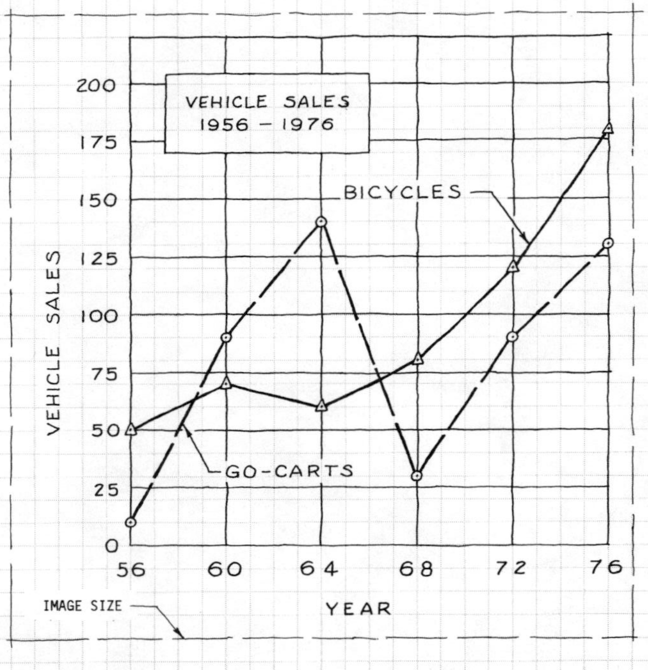

Graphics & Geometry	NAME:	MIN.	GRADE	
©	FILE: SEC: DATE:			

BREAK-EVEN GRAPH

CONSTRUCT A BREAK-EVEN GRAPH FOR MANUFACTURING THE UNIVERSAL MICROSTOP. A BREAK DOWN OF THE COST TO MANUFACTURE THE MICROSTOP IS SHOWN IN THE TABLE TO THE RIGHT. THE DEVELOPMENT COST CONSISTING OF ENGINEERING, PROTOTYPE CONSTRUCTION, AND TESTING WAS $20,000. IF YOU WISH TO BREAK EVEN WHEN 1000 ARE SOLD, WHAT MUST BE CHARGED FOR EACH MICRO-STOP? ASSUMING THAT 2400 UNITS WILL BE SOLD PLOT THE DATA AND ANSWER QUESTIONS 1 - 4.

UNIVERSAL MICROSTOP	
MANUFACTURING COSTS PER UNIT	
ITEM	COST
MATERIAL	$ 4.00
LABOR	15.00
TOOLS	3.00
TOTAL	$22.00
OVERHEAD COSTS PER UNIT	$13.00

1. UNIT COST (MANUFACTURING COST + OVERHEAD) = _
2. PRICE REQUIRED TO BREAK EVEN AT 1000 _
3. LOSS IF ONLY 600 ARE SOLD _
4. PROFIT IF 2400 UNITS ARE SOLD _

160

THOUSANDS OF DOLLARS

0

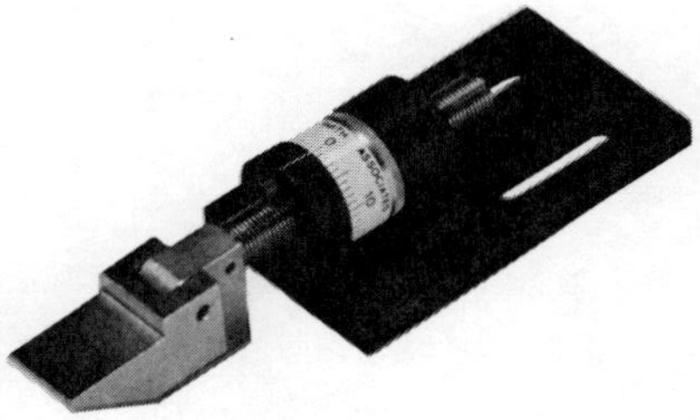

0

24

UNITS - HUNDREDS

COURTESY OF UNIVERSAL MICROSTOP

JTD

Graphics & Geometry ©

NAME:

FILE: SEC: DATE:

MIN. GRADE

91

Graphics & Geometry	NAME FILE SEC DATE	MIN.	GRADE	

GRAPH

CORROSION COSTS INDUSTRY MILLIONS OF DOLLARS EVERY YEAR. ALLOYS CAN BE USED EFFECTIVELY TO DECREASE OR ELIMINATE CORROSION. THE CORROSION RATES IN THE TABLE TO THE RIGHT ARE FOR COPPER-NICKEL ALLOYS IN A SALT SOLUTION. CORROSION DATA ARE RECORDED IN MILLIGRAMS OF METAL LOST PER SQUARE DECIMETER PER DAY (MDD).

PLOT THE DATA GIVEN IN TABLE I AND FIND THE AVERAGE CORROSION RATE FOR EACH ALLOY COMPOSITION. MARK THE AVERAGE VALUE LIGHTLY ON THE GRAPH AND DRAW THE SINGLE BEST FIT CURVE THROUGH THE AVERAGE VALUES. (SEE FIGURE I FOR AN EXAMPLE.) TO COMPLETE THE GRAPH LETTER AN APPROPRIATE TITLE IN THE TITLE SPACE PROVIDED.

TABLE I. CORROSION RATES FOR A SERIES OF COPPER-NICKEL ALLOYS IMMERSED IN A SALT SOLUTION.

ALLOY COMPOSITION	SAMPLE NUMBER	WEIGHT OF METAL LOST (MILLIGRAMS/SQ. DM./DAY, MDD)
100% CU	1	102
	2	96
80% CU-20% NI	1	12
	2	10
60% CU-40% NI	1	7.3
	2	4.9
40% CU-60% NI	1	5.8
	2	4.7
20% CU-80% NI	1	6.5
	2	4.6
100% NI	1	7.8
	2	5.5

CORROSION RATE, MDD

100

10

3

0 20 40 60 80 100

WEIGHT PER CENT NICKEL

AVERAGE VALUE

USE CURVES AND SYMBOLS THIS SIZE

FIGURE 1.

COURTESY OF JTC

JTD

JTD

METRIC SCALES

Remove this page and fold it longwise along the desired scale for making metric measurements.

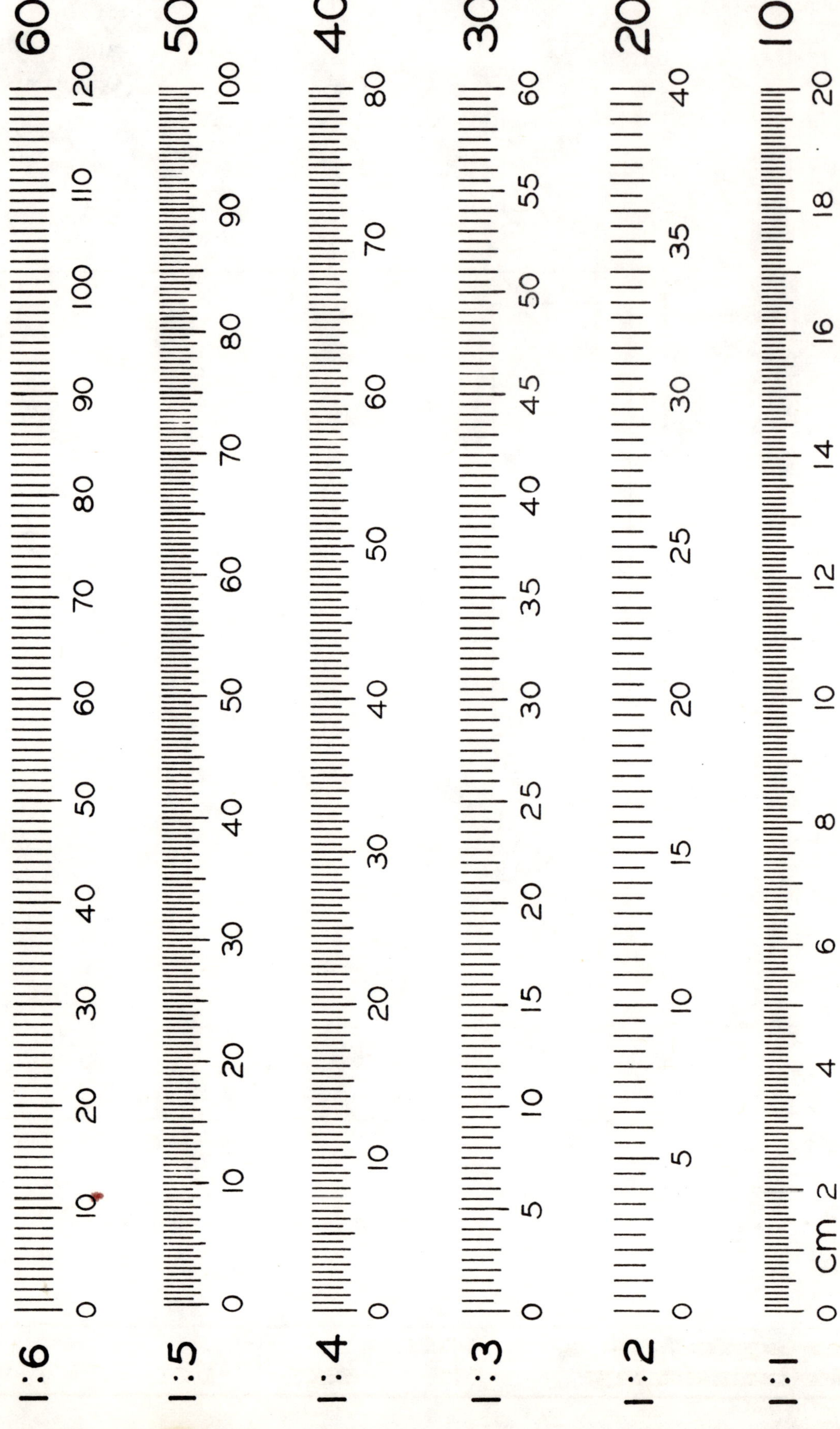

DATA ANALYSIS

A FR70-14 SET OF RADIAL TIRES HAS BEEN TESTED ON A WET SURFACE AND THE FOLLOWING READINGS HAVE BEEN RECORDED IN THE TABLES BELOW.

1. PLOT THE THREE CURVES OF CORNERING FORCE Fy VERSUS STEERING ANGLE α FOR EACH SPEED OF 20, 40, AND 60 MPH.
2. GRAPHICALLY SUBSTRACT THE 40 MPH CURVE FROM THE 20 MPH CURVE, AND THE 60 MPH CURVE FROM THE 40 MPH CURVE AND SHOW THE DECREASE OF THE CORNERING FORCE VERSUS SPEED USING THE ORDINATE SCALE AT THE RIGHT OF GRAPH.

COURTESY MTS

QUESTIONS

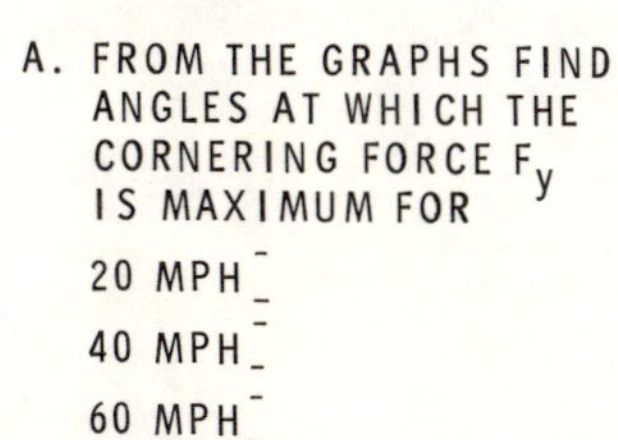

A. FROM THE GRAPHS FIND ANGLES AT WHICH THE CORNERING FORCE F_y IS MAXIMUM FOR

20 MPH___

40 MPH___

60 MPH___

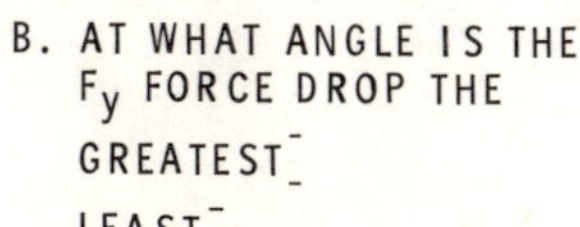

B. AT WHAT ANGLE IS THE F_y FORCE DROP THE

GREATEST___

LEAST___

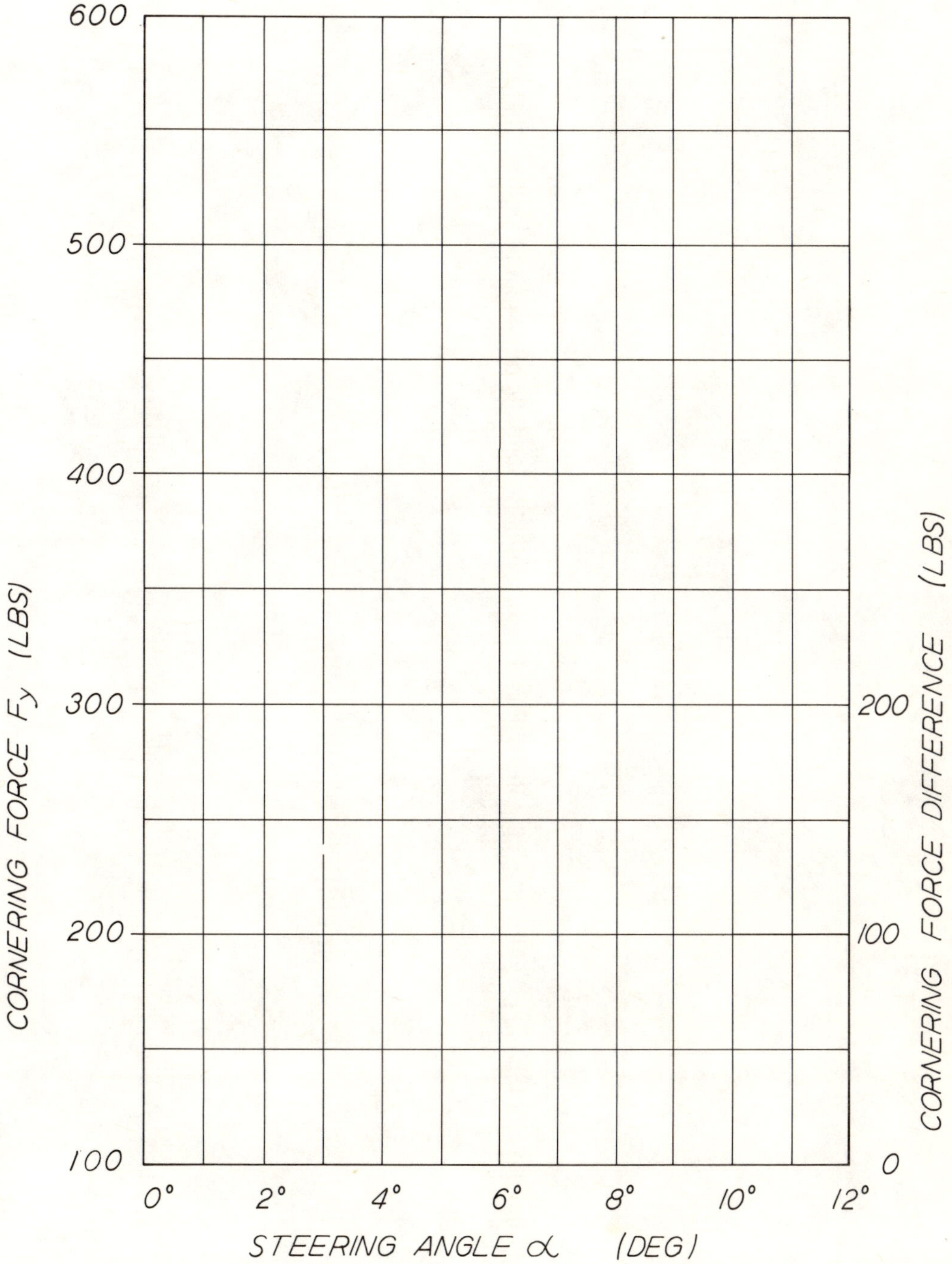

DATA

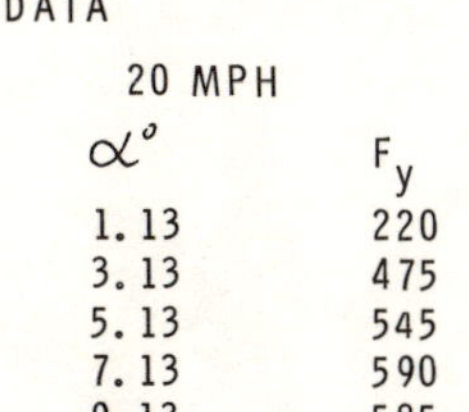

20 MPH

$\alpha°$	F_y
1.13	220
3.13	475
5.13	545
7.13	590
9.13	585
11.1	570

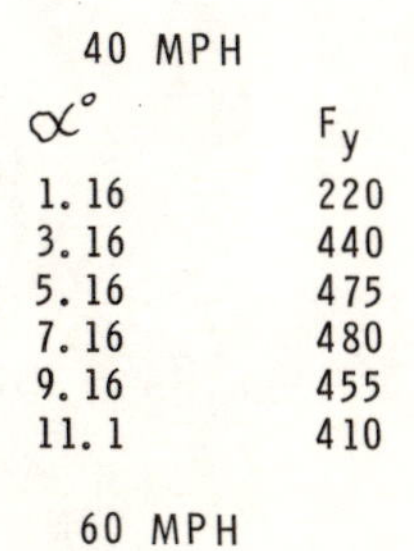

40 MPH

$\alpha°$	F_y
1.16	220
3.16	440
5.16	475
7.16	480
9.16	455
11.1	410

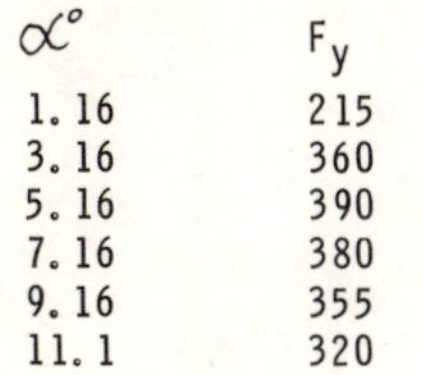

60 MPH

$\alpha°$	F_y
1.16	215
3.16	360
5.16	390
7.16	380
9.16	355
11.1	320

CORNERING FORCE F_y (LBS): 600, 500, 400, 300, 200, 100

CORNERING FORCE DIFFERENCE (LBS): 200, 100, 0

STEERING ANGLE α (DEG): 0°, 2°, 4°, 6°, 8°, 10°, 12°

GR

Graphics & Geometry ©	NAME___ FILE___ SEC___ DATE___	MIN.	GRADE	93

METRIC SCALES

Remove this page and fold it longwise along the desired scale for making metric measurements.

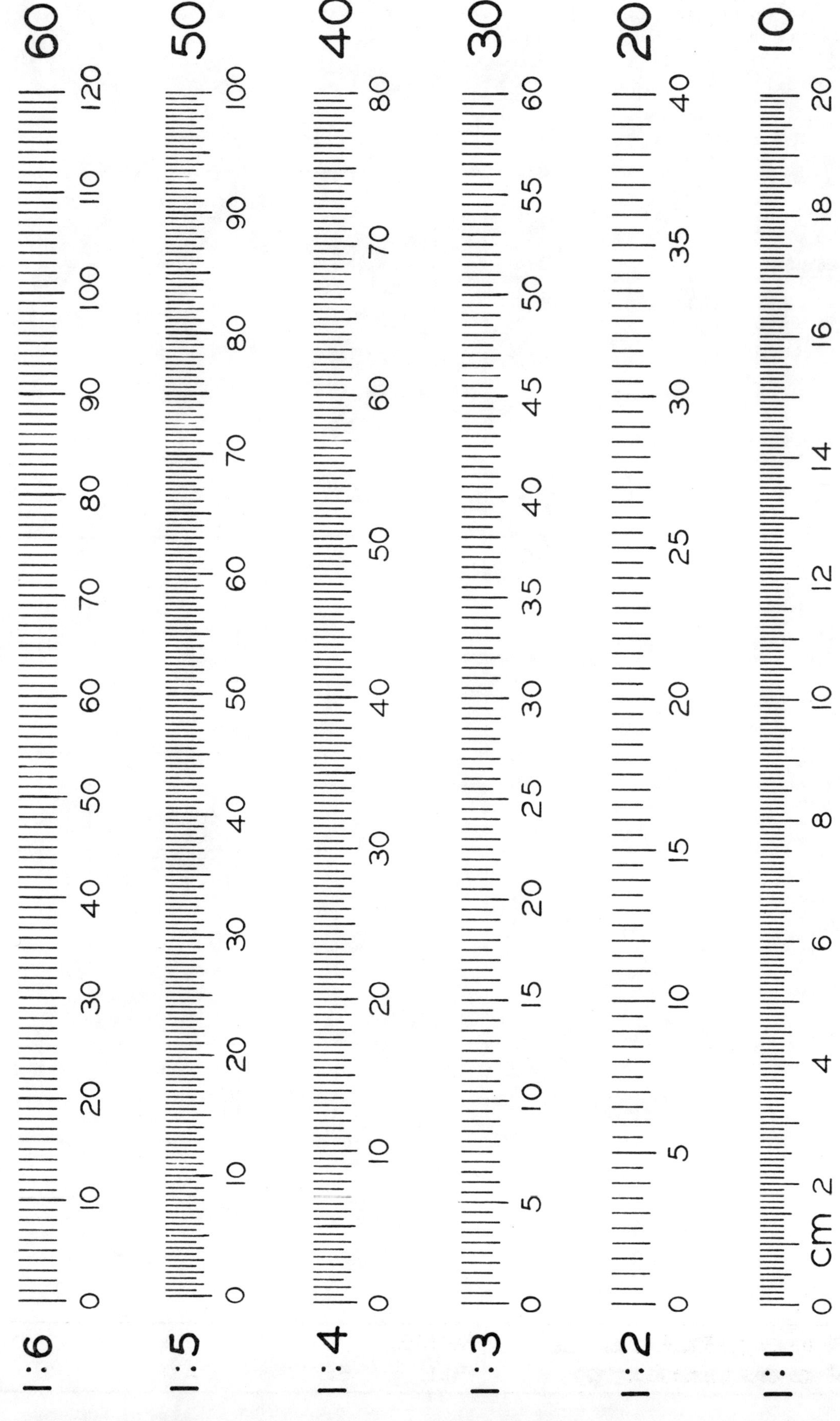

EMPIRICAL EQUATIONS

EQUATIONS DERIVED FROM EXPERIMENTAL DATA ARE CALLED EMPIRICAL EQUATIONS. THESE EQUATIONS CAN BE FOUND WHEN DATA PLOTS AS A STRAIGHT LINE ON ONE OF THE THREE GRAPHS SHOWN HERE. PLOT THE DATA FOR EACH PROBLEM ON THE GRAPH THAT WILL GIVE A STRAIGHT LINE. SINCE THE DATA IS EXPERIMENTAL, YOU MUST FIT THE CURVE TO GIVE THE BEST STRAIGHT LINE. DETERMINE THE EQUATIONS FOR EACH PROBLEM USING BASE 10 LOGARITHMS. LABEL EACH X- AND Y- AXIS.

PROB 1: YOUR MANAGEMENT PEOPLE HAVE ESTIMATED THE COST PER ITEM FOR MANUFACTURING VARIOUS QUANTITIES OF A PRODUCT. WHAT IS THE EQUATION OF THIS DATA? WHAT IS THE PERCENT INCREASE PER UNIT COST FOR 150 ITEMS COMPARED WITH THE UNIT COST AT 500?

PROB 2: A SHUTTLE MOVES LARGE PARTS ON AN ASSEMBLY LINE AT VARIOUS SPEEDS THAT HAVE BEEN MEASURED DURING A SEVEN-MINUTE INTERVAL. WHAT IS THE EQUATION OF THIS DATA?

PROB 3: IT HAS BEEN DETERMINED THAT THERE IS A RELATIONSHIP BETWEEN THE SPEED OF A FLUID IN A PIPE SYSTEM AND THE OUTPUT OF THE FLUID IN GALLONS PER MINUTE. WHAT IS THE EQUATION OF THIS DATA?

PROB 1		PROB 2		PROB 3	
NUMBER	COST/ITEM	MIN	FT/MIN	GAL/MIN	FT/SEC
X	Y	X	Y	X	Y
0	—	0	80	0	-
10	4.20	1	71	0.1	5
20	3.25	2	58	0.4	11
30	2.50	3	46	0.6	15
40	2.00	4	38	1	20
50	1.40	5	25	2	30
60	1.10	6	13	4	45
70	.82	7	2	6	58

EQUATION FORMS

ST. LINE	FORM	M = SLOPE	B = INTERCEPT
RECTILINEAR	$Y = MX + B$	$\frac{Y_2 - Y_1}{X_2 - X_1}$	WHERE $X = 0$
SEMILOG	$Y = BM^X$	$\frac{\log Y_2 - \log Y_1}{X_2 - X_1}$	WHERE $X = 0$
LOG-LOG	$Y = BX^M$	$\frac{\log Y_2 - \log Y_1}{\log X_2 - \log X_1}$	WHERE $X = 1$

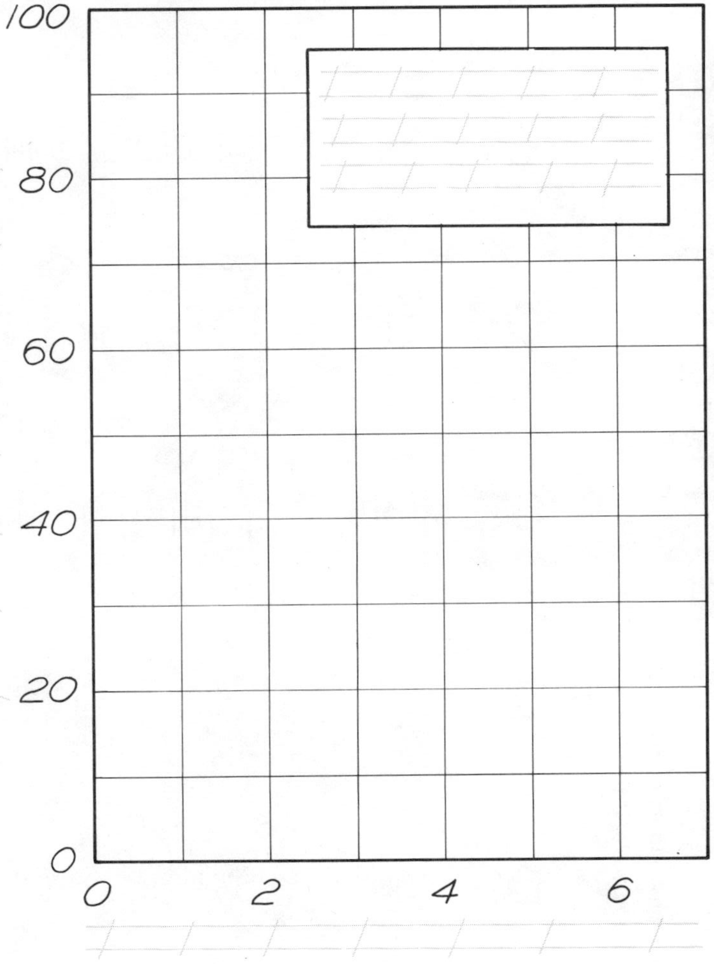

THE STRAIGHT LINE IS PROB.

EQUATION FORM: $Y = MX + B$

$M = \frac{Y_2 - Y_1}{X_2 - X_1} = \frac{\quad - \quad}{\quad - \quad} =$

$B =$

EQUATION: $Y =$

Graphics & Geometry ©

NAME FILE SEC DATE MIN. GRADE

JHC

METRIC SCALES

Remove this page and fold it longwise along the desired scale for making metric measurements.

EMPIRICAL EQUATIONS

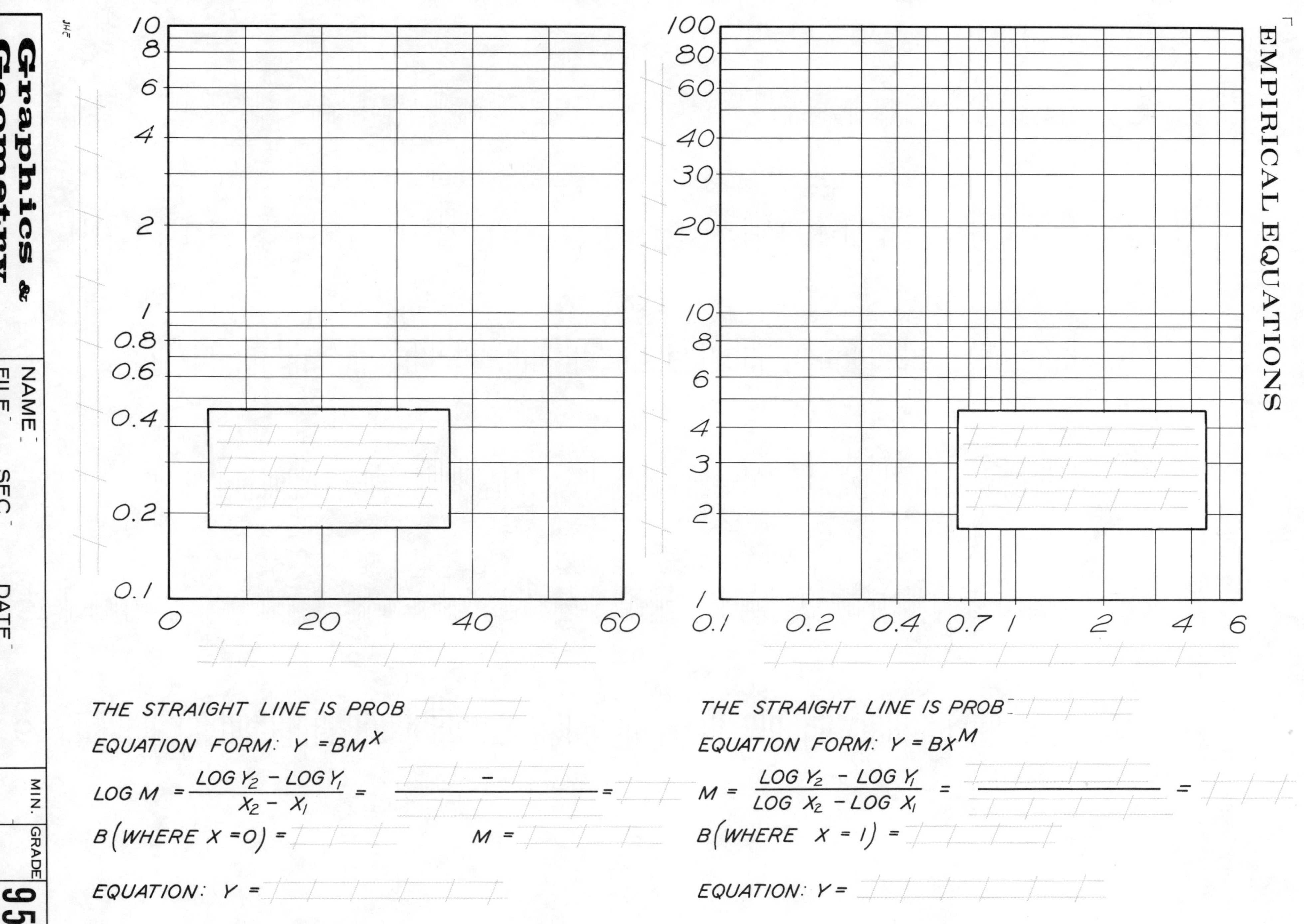

THE STRAIGHT LINE IS PROB

EQUATION FORM: $Y = BM^X$

$\log M = \frac{\log Y_2 - \log Y_1}{X_2 - X_1} = \frac{\quad - \quad}{\qquad} =$

B (WHERE X = 0) = M =

EQUATION: Y =

THE STRAIGHT LINE IS PROB

EQUATION FORM: $Y = BX^M$

$M = \frac{\log Y_2 - \log Y_1}{\log X_2 - \log X_1} = \frac{\qquad}{\qquad} =$

B (WHERE X = 1) =

EQUATION: Y =

JHE

Graphics & Geometry ©	NAME FILE SEC DATE	MIN.	GRADE	95

METRIC SCALES

Remove this page and fold it longwise along the desired scale for making metric measurements.

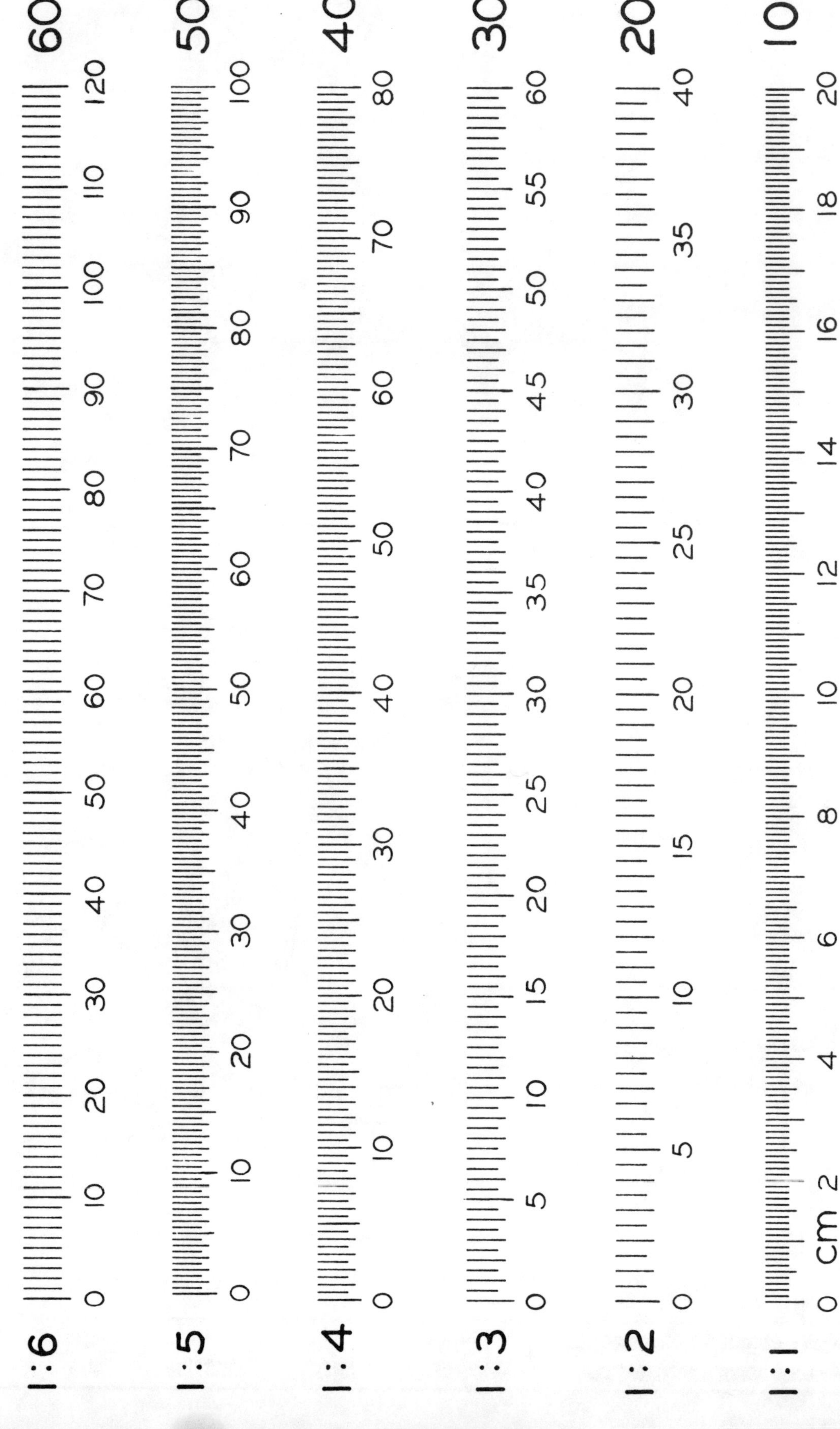

NOMOGRAMS

A CONTRACTOR MUST ALLOW A MARGIN OF PROFIT WHEN HE BIDS ON A PROJECT. ASSUME THAT YOU MUST HAVE A MARGIN OF 60% ON LABOR COSTS AND 40% ON MATERIALS COSTS WHEN BIDDING. THIS CAN BE WRITTEN AS AN EQUATION

$$BC = 1.6\ (L) + 1.4\ (M)$$

WHERE BC = BID COST; L = LABOR COST; M = MATERIAL COST.

CONSTRUCT A PARALLEL SCALE CHART FOR PREPARING YOUR BIDS. REFER TO THE KEY BELOW. USE 20-SCALE FOR MATERIAL SCALE AND 40-SCALE FOR LABOR SCALE.

COURTESY BUREAU OF RECLAMATION
U.S. DEPARTMENT OF THE INTERIOR

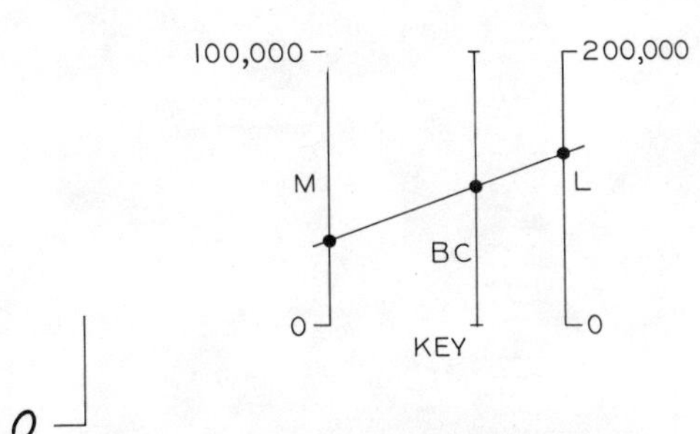

0

0

1. WHEN ACTUAL LABOR COSTS ARE $125,000 AND MATERIALS COSTS ARE $50,000, WHAT WILL BE YOUR BID (BC)?
2. FOR A TOTAL BID (BC) OF $200,000 AND A MATERIALS COST OF $35,000, WHAT WOULD BE THE ACTUAL COST OF LABOR?
3. FOR A TOTAL BID (BC) OF $300,000 AND A MATERIALS COST OF $48,000, WHAT WOULD BE THE ACTUAL COST OF LABOR?

JHE

METRIC SCALES

Remove this page and fold it longwise along the desired scale for making metric measurements.

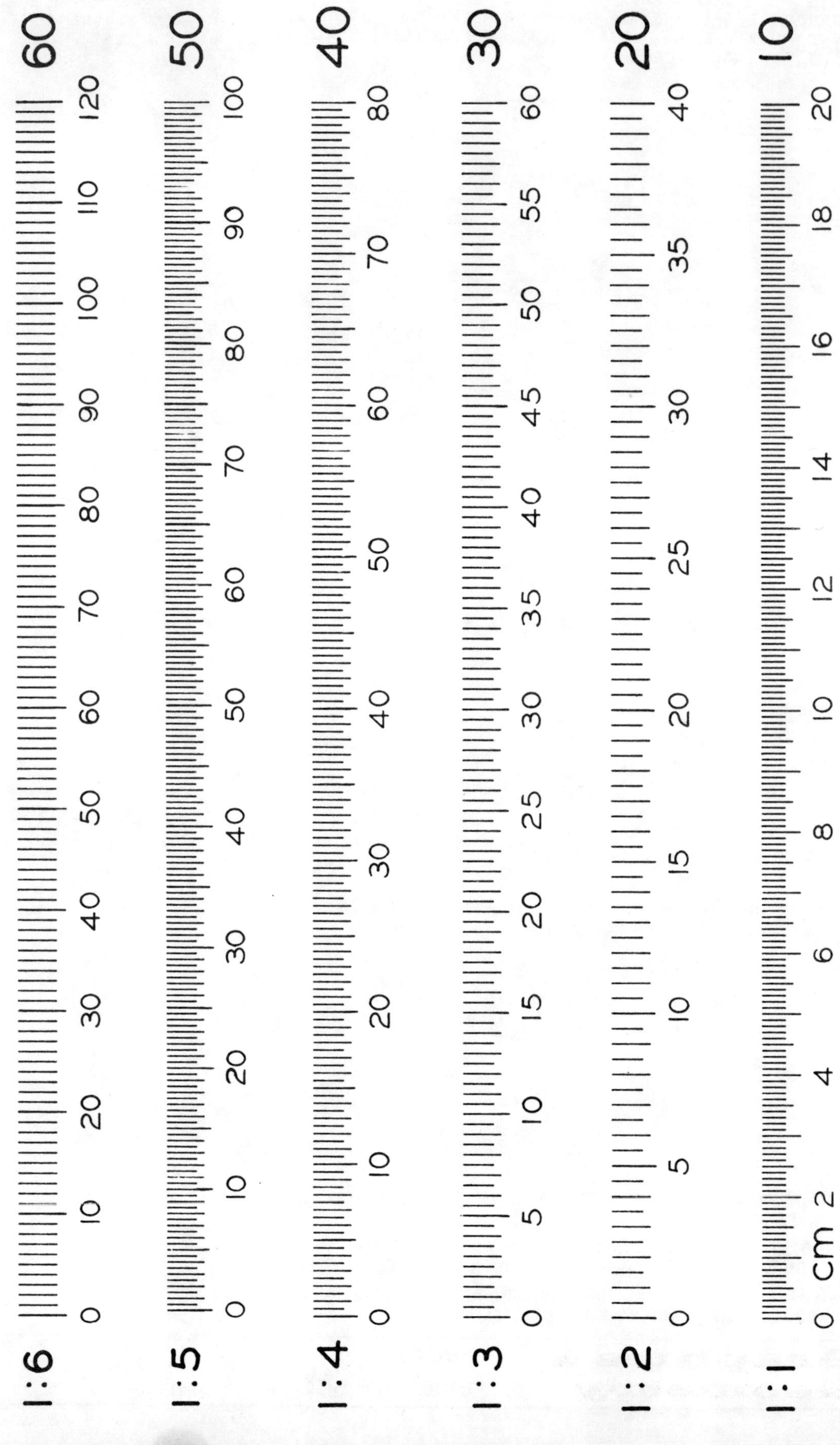

NOMOGRAMS

VARIOUS SIZES OF ELECTRICAL WIRES HAVE DIFFERENT RESISTANCES (IN OHMS) THAT ARE RELATED TO THE CROSS SECTION OF THE WIRE. THIS RESISTANCE CAUSES A LOSS IN VOLTS WHEN ELECTRICITY IS TRANSMITTED THROUGH LENGTHS OF WIRE. THIS LOSS CAN BE DETERMINED BY OHM'S LAW E = IR, WHERE E IS VOLTS, I IS AMPERES AND R IS RESISTANCE IN OHMS.

THE N-CHART BELOW IS PARTIALLY COMPLETED. THE VOLTAGE DROP IS CALIBRATED FROM 0 TO 10. THE RESISTANCE SCALE IS GIVEN FROM 0 TO 3.0 OHMS. NOTICE THAT WIRE SIZES OF BARE COPPER WIRE ARE GIVEN ON THIS SCALE TO CORRESPOND TO THE AMOUNT OF RESISTANCE IN VARIOUS SIZES OF WIRE OF 1000 FT. LENGTHS. FOR EXAMPLE, A #10 WIRE WILL HAVE A RESISTANCE OF 1.018 OHMS FOR A LENGTH OF 1000 FEET ACCORDING TO THE NATIONAL ELECTRIC CODE.

REQUIRED

COMPLETE THE N-CHART BY CALIBRATING THE DIAGONAL SCALE. LABEL EVEN UNITS ALONG THIS SCALE. ANSWER QUESTIONS BELOW.

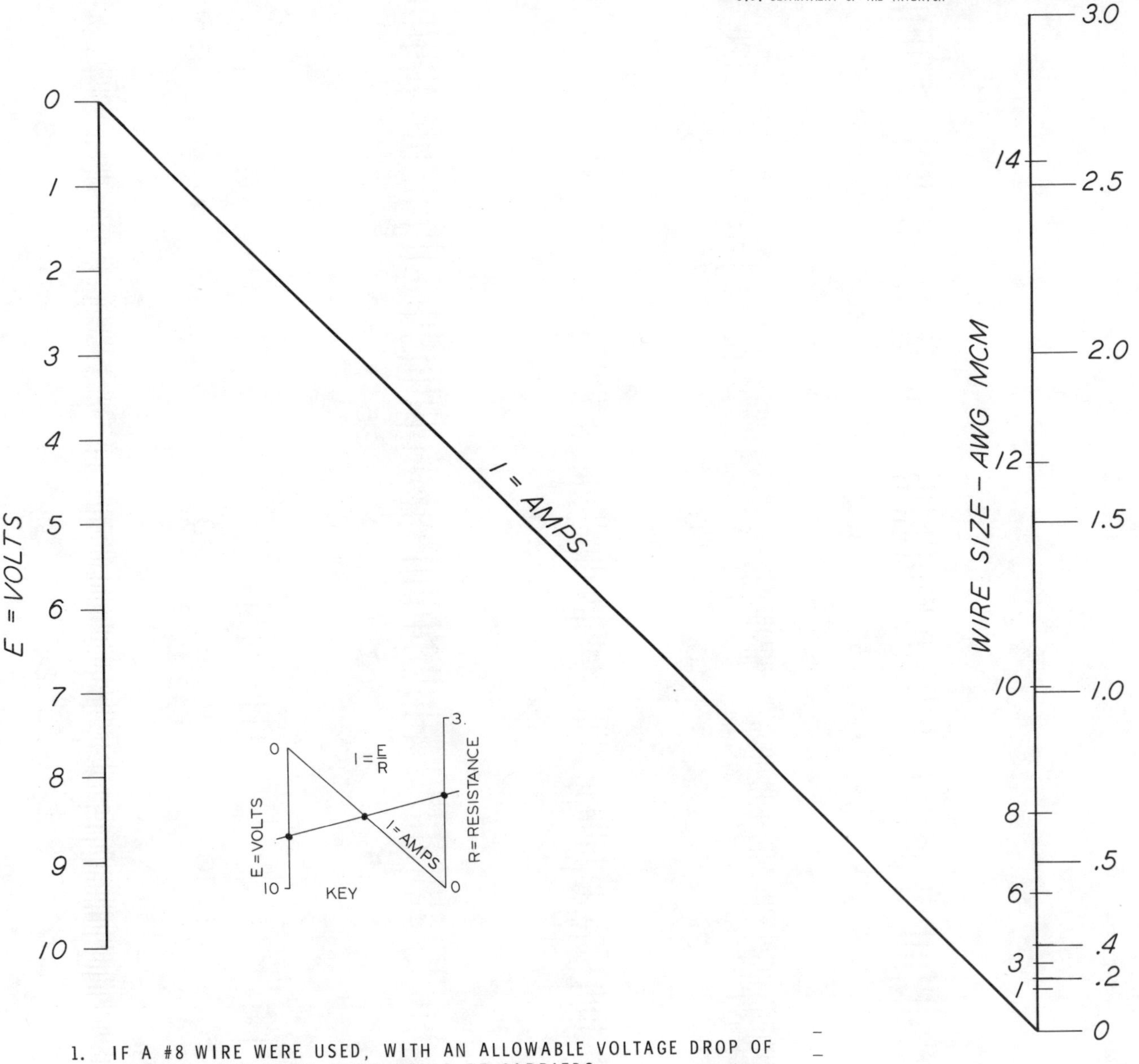

COURTESY BUREAU OF RECLAMATION
U.S. DEPARTMENT OF THE INTERIOR

E = VOLTS: 0, 1, 2, 3, 4, 5, 6, 7, 8, 9, 10

I = AMPS

WIRE SIZE - AWG MCM: 14, 12, 10, 8, 6, 3, 1

R = OHMS: 3.0, 2.5, 2.0, 1.5, 1.0, .5, .4, .2, 0

KEY: E = VOLTS (0, 10); $I = \frac{E}{R}$; I = AMPS; R = RESISTANCE (3., 0)

1. IF A #8 WIRE WERE USED, WITH AN ALLOWABLE VOLTAGE DROP OF 5 VOLTS, HOW MANY AMPS WOULD BE CARRIED?
2. WITH AN ALLOWABLE VOLTAGE LOSS OF 8 VOLTS, WHAT SIZE WIRE WOULD CARRY 10 AMPS?
3. IF 6 AMPS ARE TO BE CARRIED WITH A RESISTANCE OF 1.5 OHMS, WHAT WOULD BE THE VOLTAGE DROP?

JHE

METRIC SCALES

Remove this page and fold it longwise along the desired scale for making metric measurements.

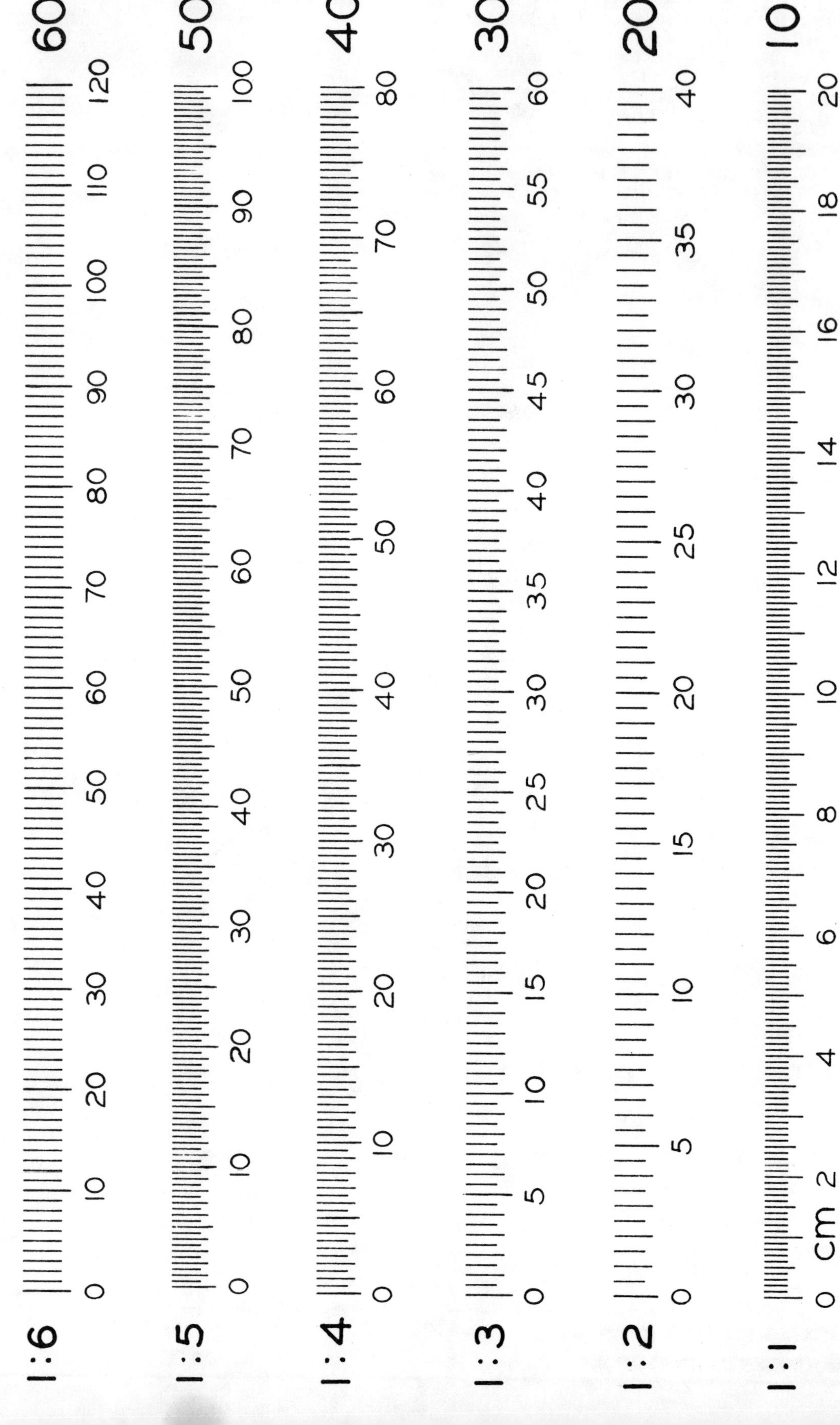

GRAPHICAL DIFFERENTIATION

A CAM HAS BEEN DESIGNED THAT WILL CAUSE A FOLLOWER-RISE AT EACH INTERVAL OF ITS CONSTANT REVOLUTION AS PLOTTED IN THE GRAPH. CONNECT THESE POINTS WITH A SMOOTH CURVE. FIND THE VELOCITY CURVE OF THIS FOLLOWER BY GRAPHICAL DIFFERENTIATION. LABEL THE AXES AND GIVE A TITLE IN THE BLOCK PROVIDED. ANSWER THE

AT WHAT TIME IS THE VELOCITY ZERO?

WHAT IS THE MAXIMUM VELOCITY?

AT WHAT TIME IS THE VELOCITY MAXIMUM?

WHAT IS THE RPM OF THE CAM?

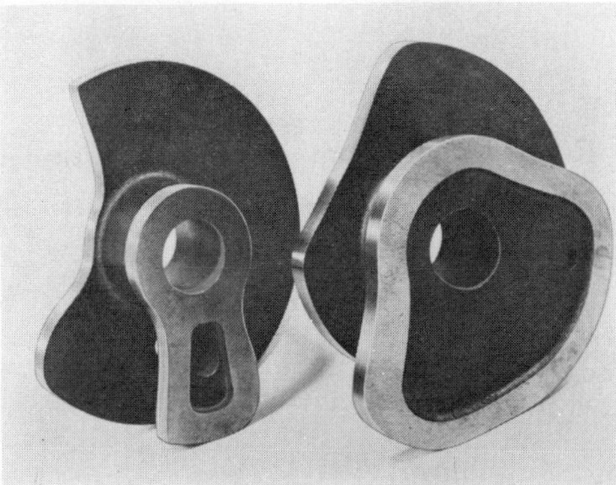

COURTESY FERGUSON MACHINE COMPANY

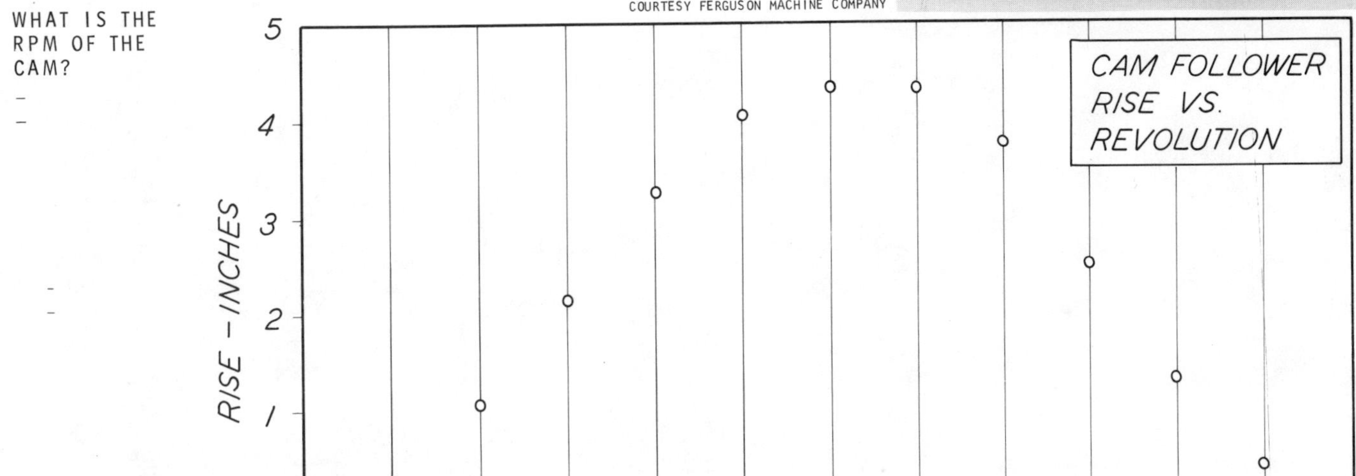

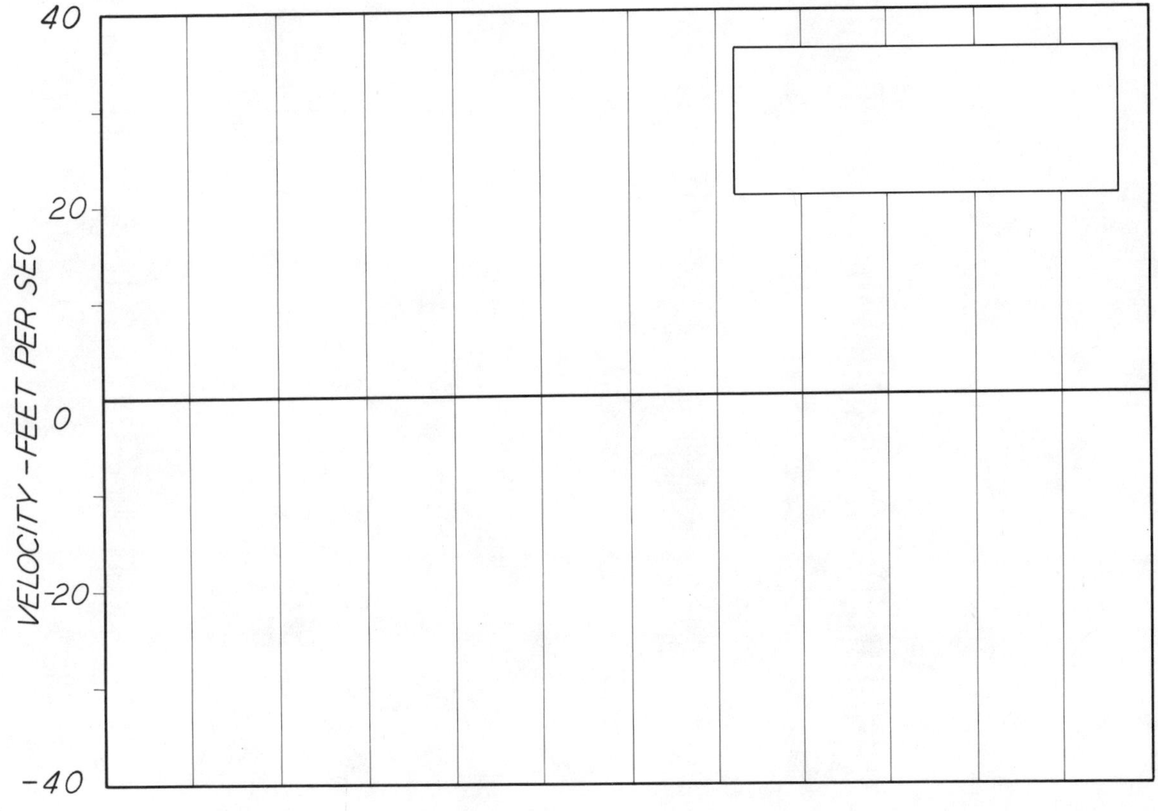

METRIC SCALES

Remove this page and fold it longwise along the desired scale for making metric measurements.

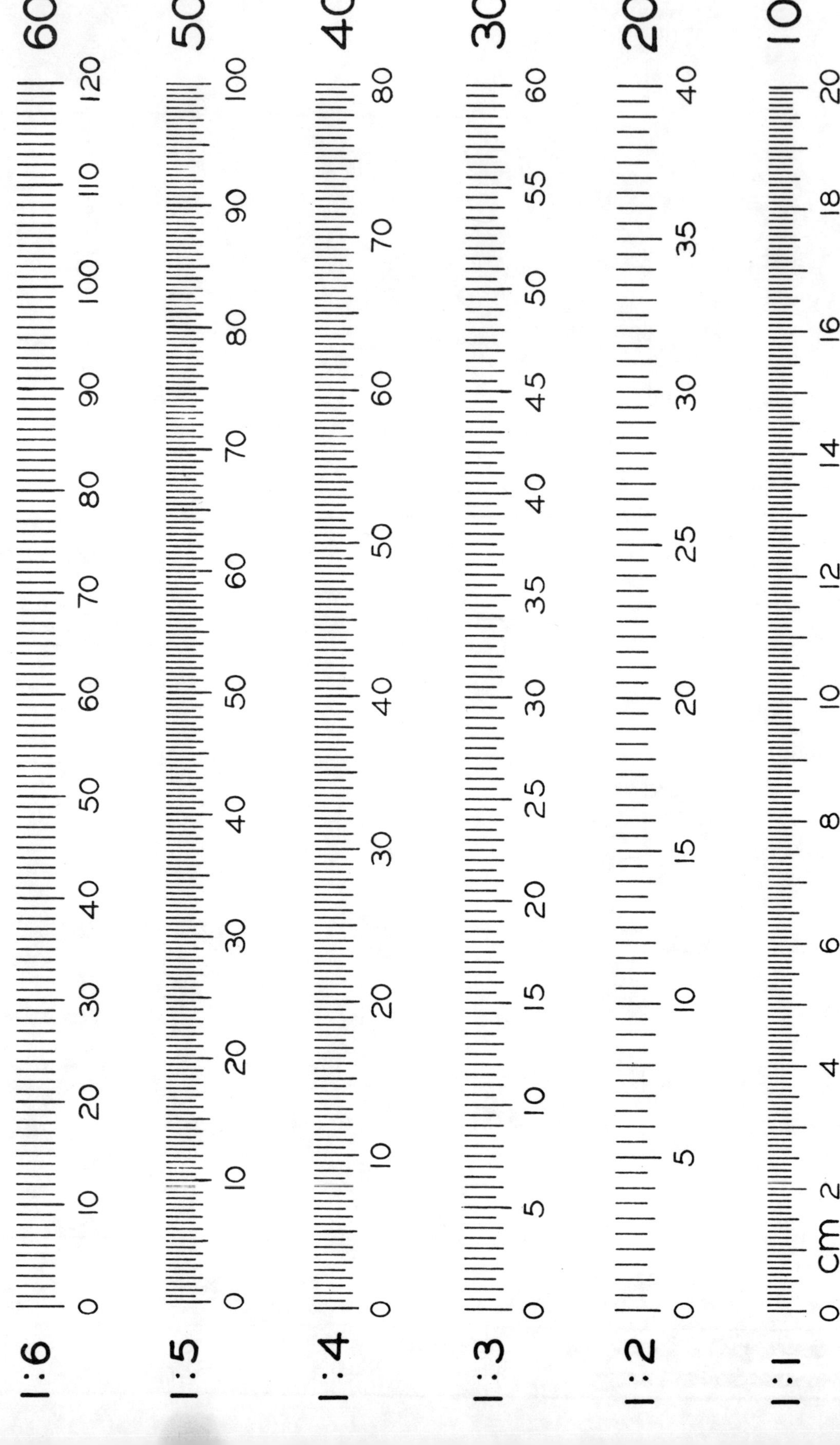

INTEGRATION

A SKETCH OF AN AIR-LOADING CURVE FOR A WING IS SHOWN BELOW. THIS NEGATIVE LOAD IS PLOTTED IN THE LOWER GRAPH. FIND THE SHEAR CURVE BY INTEGRATING THE GIVEN DATA FROM 0 TO 50 FEET.

1 WHAT IS THE MAXIMUM SHEAR OF THE WING? ___

2 WHAT IS THE SHEAR 30 FEET FROM THE WING TIP? ___

3 HOW FAR FROM THE WING TIP IS SHEAR THE GREATEST? ___

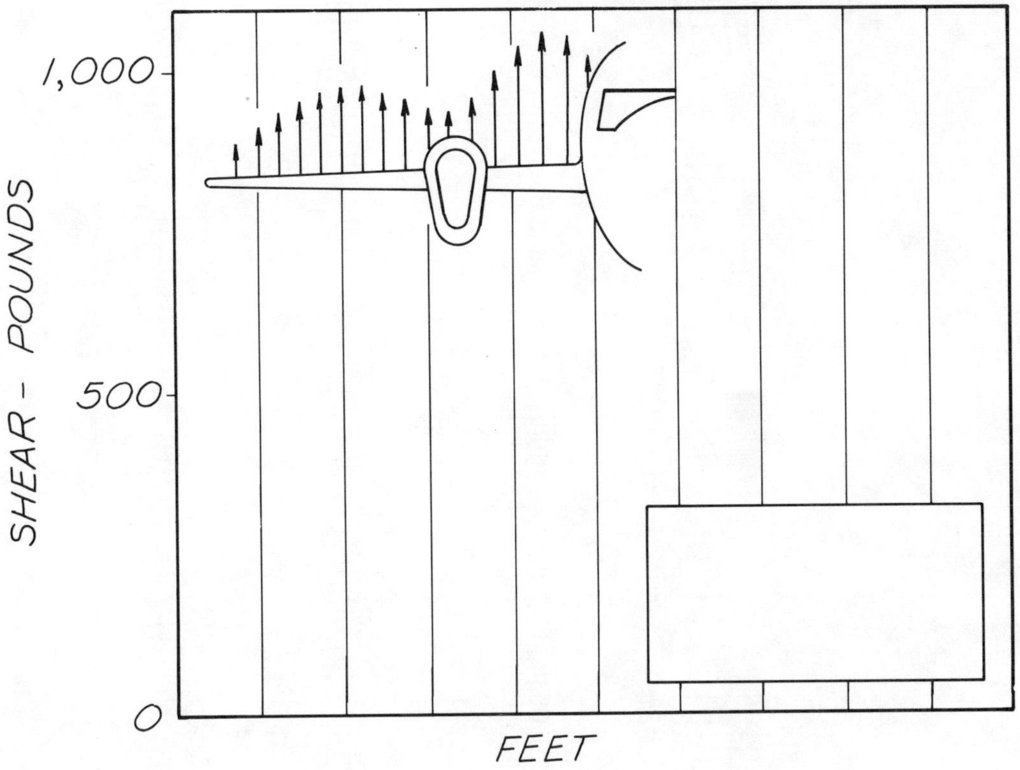

SHEAR - POUNDS

1,000

500

0

FEET

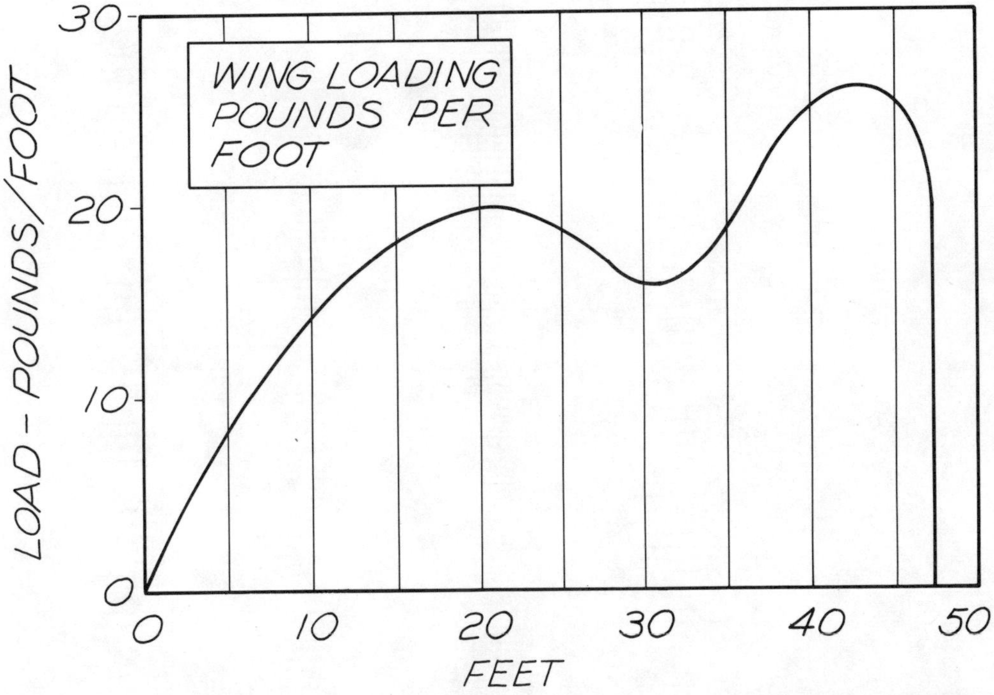

WING LOADING POUNDS PER FOOT

LOAD - POUNDS/FOOT

30

20

10

0

0 10 20 30 40 50

FEET

JHE

METRIC SCALES

Remove this page and fold it longwise along the desired scale for making metric measurements.

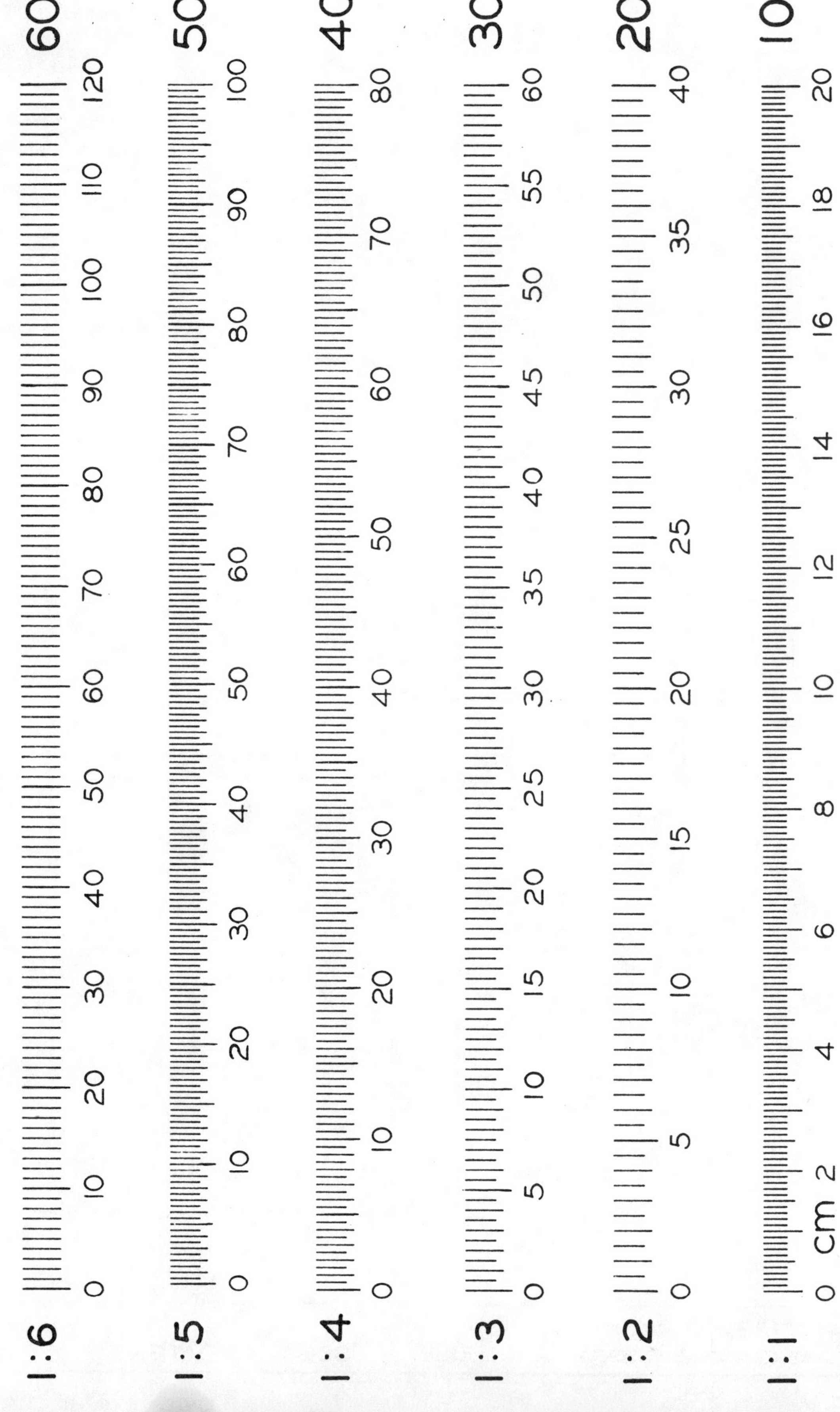

THREAD NOTES

COMPLETE THE THREAD NOTES FOR PROBLEMS 1, 2, & 3 BY MEASURING THE DIAMETERS AND USING THE THREAD TABLES IN YOUR TEXTBOOK. USE THE FOLLOWING SPECIFICATIONS:

PROBLEM	SERIES	FIT	OTHER
1	UNC	2	L.H.
2	UNF	2	--
3	NC	1	DOUBLE

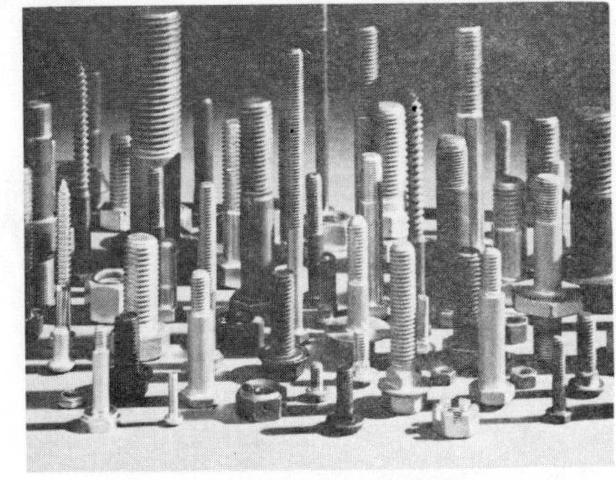

RUSSELL, BURDSALL & WARD BOLT AND NUT COMPANY

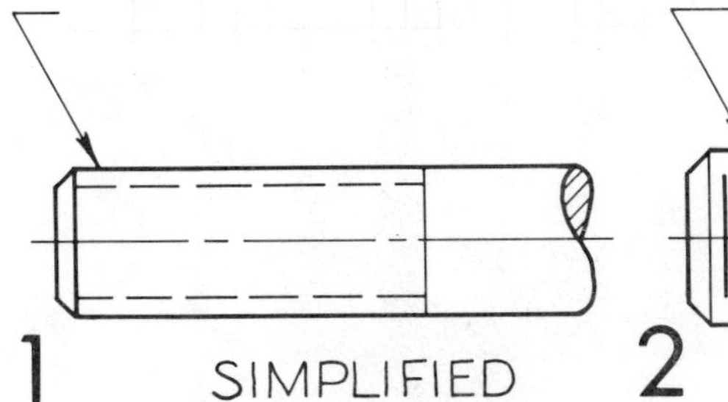

1 SIMPLIFIED

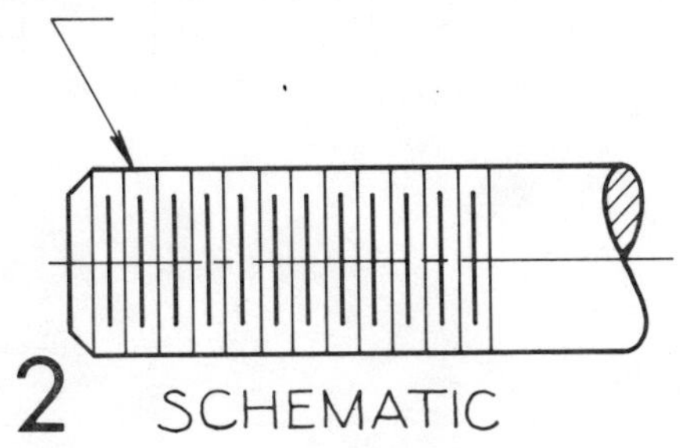

2 SCHEMATIC

COMPLETE THE THREAD NOTES ATTACHED TO THE LEADERS TO SPECIFY THE INTERNAL THREADS SHOWN IN THE VIEWS AND SECTIONS BELOW. MEASURE THE MAJOR DIAMETERS AND USE YOUR THREAD TABLES AND THE SPECIFICATIONS BELOW:

PROB.	SERIES	FIT	OTHER
4	UNF	2	L.H., DOUBLE
5	UNC	2	R.H., SINGLE
6	NC	1	R.H., TRIPLE
7	UNEF	2	L.H., SINGLE

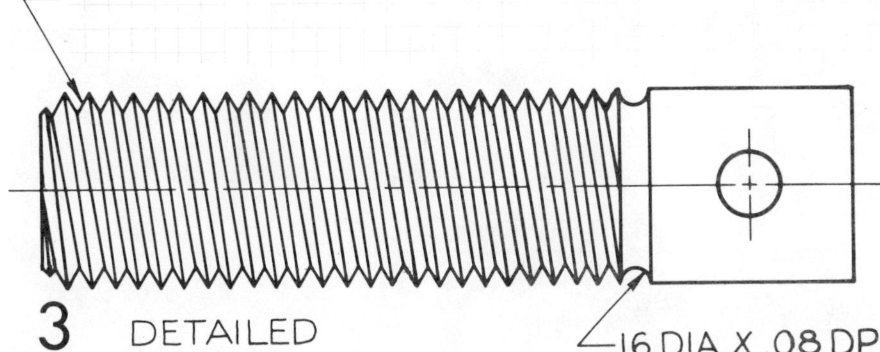

3 DETAILED

4

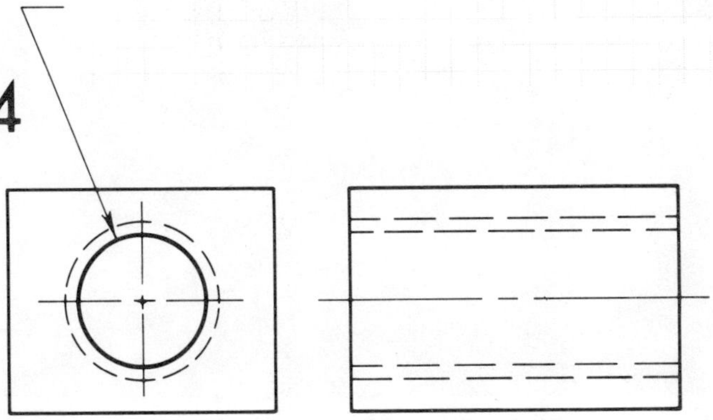

SIMPLIFIED VIEW

5

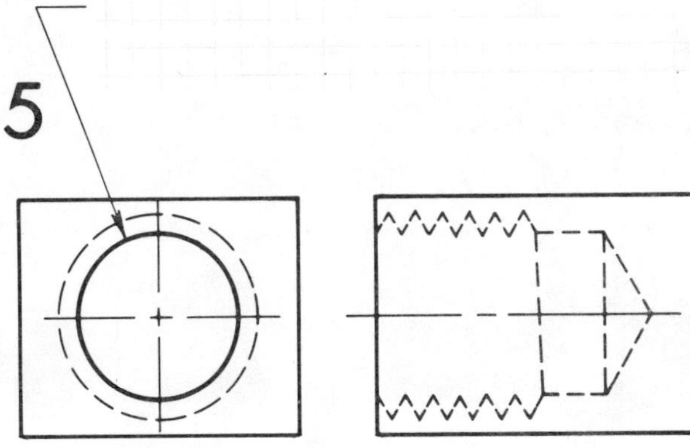

DETAILED VIEW

6

SIMPLIFIED SECTION

7

SCHEMATIC SECTION

JHE

Graphics & Geometry ©

NAME
FILE SEC DATE

MIN. GRADE **100**

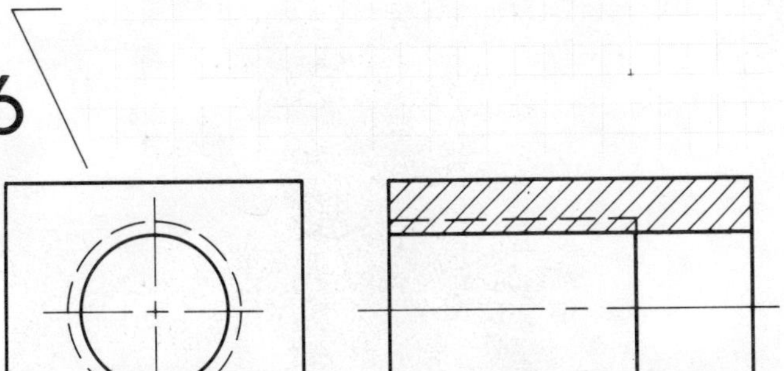

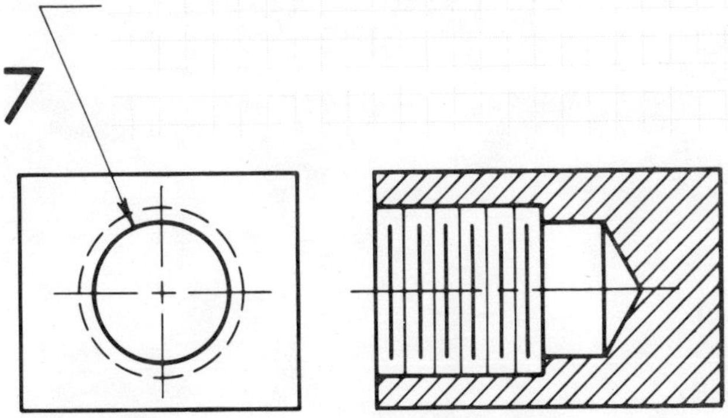

1

MEASURE THE SCREWS BELOW, USE YOUR THREAD TABLES, AND GIVE NOTES TO INDICATE THE TYPES OF THREADS AND SCREWS. DRAW THREADS ON EACH USING SCHEMATIC SYMBOLS AT A & B AND SIMPLIFIED SYMBOLS AT C. EACH IS A UNC SERIES WITH A CLASS 2 FIT. DRAW THREADS FROM THE END OF EACH TO POINT O.

THREAD SYMBOLS

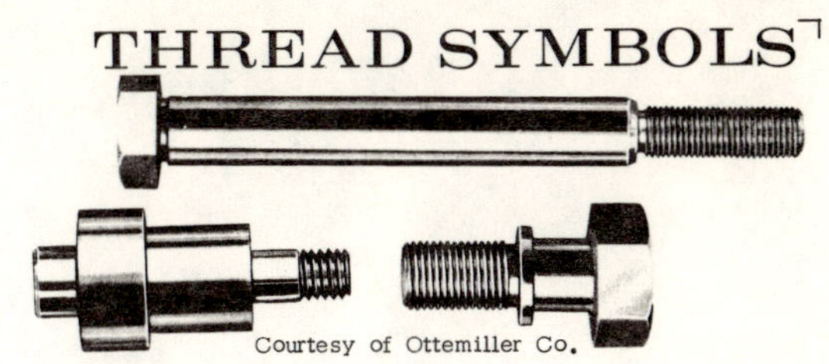

Courtesy of Ottemiller Co.

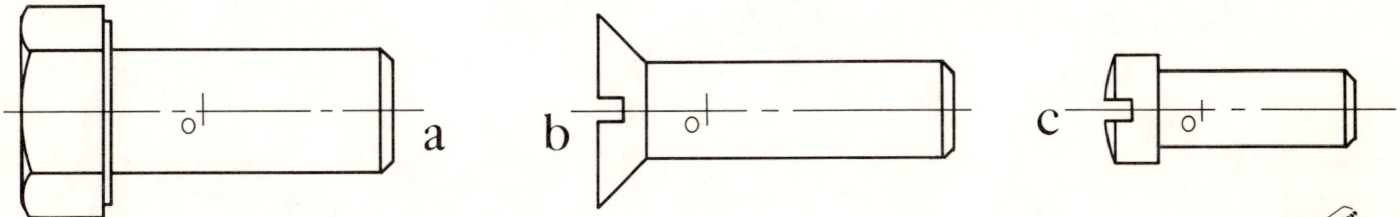

O
a
b
O
c
O

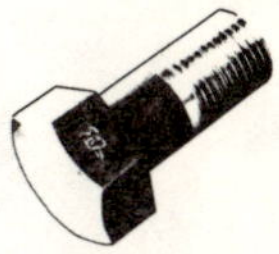

HEXAGON HEAD BOLT

2

USING CENTER, C, DRAW THE HALF-VIEW OF A HEXAGON HEAD ACROSS CORNERS. PROJECT TO THE VIEW AT THE RIGHT AND SHOW THE HEAD BETWEEN POINTS L & M. DRAW DETAILED THREAD SYMBOLS FROM N TO O. GIVE A THREAD NOTE FOR A UNC THREAD WITH A CLASS 2 FIT.

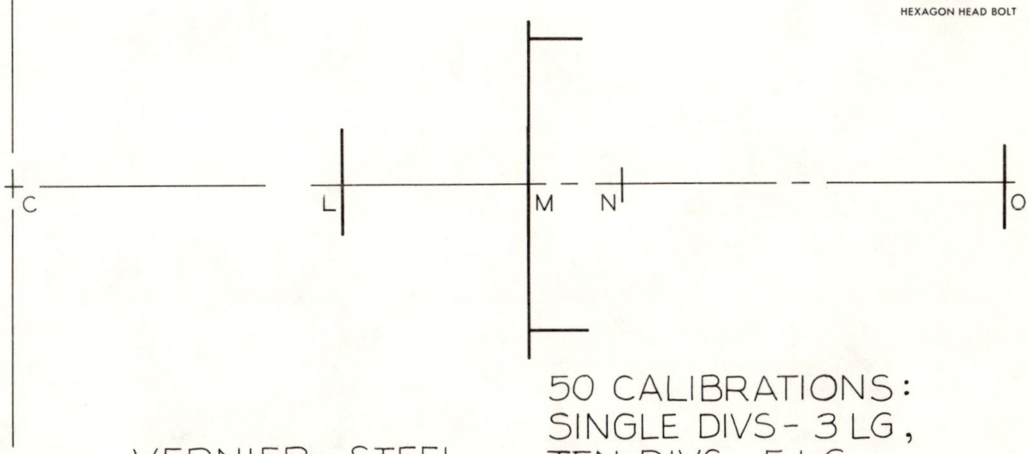

C
L
M
N
O

3

GIVE THREAD NOTES FOR THE SHAFT AND VERNIER, PARTS OF THE MICRO-STOP ASSEMBLY THAT FIT TOGETHER. USE SIMPLIFIED THREAD SYMBOLS ON THE SHAFT THAT IS THREADED ITS FULL LENGTH. SPECIFICATIONS: 1/2 - 20 UNF - 2 OR M12 X 1.25.

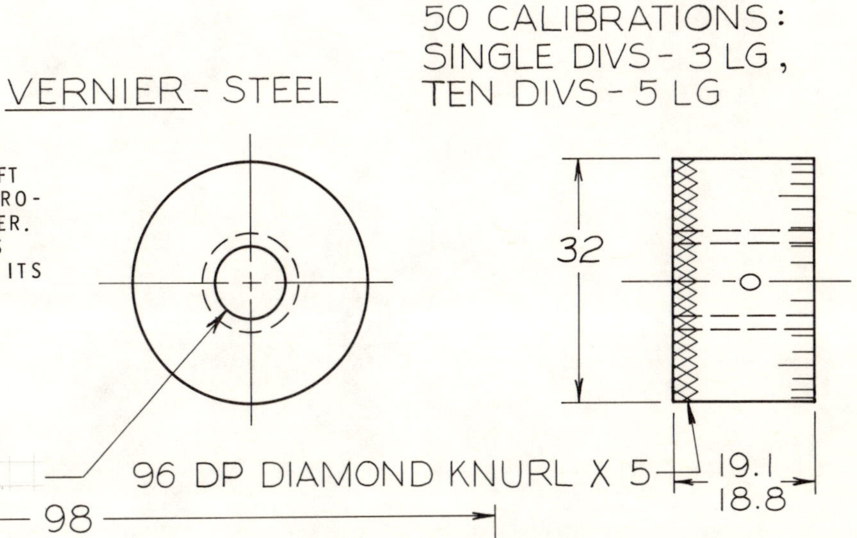

VERNIER - STEEL
50 CALIBRATIONS:
SINGLE DIVS - 3 LG,
TEN DIVS - 5 LG
32
O
96 DP DIAMOND KNURL X 5
19.1
18.8

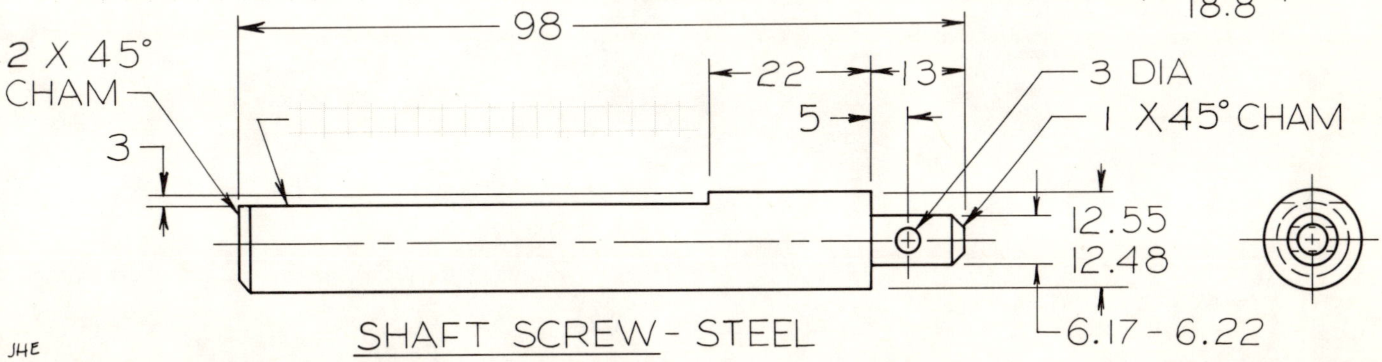

98
2 X 45°
CHAM
22
13
3 DIA
5
1 X45° CHAM
3
12.55
12.48
6.17 - 6.22
SHAFT SCREW - STEEL

JHE

JHE

Graphics & Geometry ©	NAME FILE SEC DATE	MIN.	GRADE	101

COMPLETE THE REPRESENTATION OF THE THREADS IN PROBLEMS 1 & 2 USING THE SYMBOLS SPECIFIED. MEASURE THE FULL SIZE DIAMETERS OF EACH AND COMPLETE THE METRIC THREAD NOTES FOR COARSE THREADS USING THE TABLE.

BASIC DESIGNATIONS

COARSE	FINE
M 14	M 14 X 1.5
M 14 X 2	PREFERRED USA PRACTICE

METRIC THREADS

BASIC THREAD DESIGNATIONS FOR COMMERCIAL SERIES OF ISO METRIC THREADS

Nominal Size (mm)	Pitch P (mm)	Basic Thread Designation *	Nominal Size (mm)	Pitch P (mm)	Basic Thread Designation *
1.6	0.35	M1.6	14	2 1.5	M14 M14 X 1.5
1.8	0.35	M1.8	16	2 1.5	M16 M16 X 1.5
2	0.4	M2	18	2.5 1.5	M18 M18 X 1.5
2.2	0.45	M2.2	20	2.5 1.5	M20 M20 X 1.5
2.5	0.45	M2.5	22	2.5 1.5	M22 M22 X 1.5
3	0.5	M3	24	3 2	M24 M24 X 2
3.5	0.6	M3.5	27	3 2	M27 M27 X 2
4	0.7	M4	30	3.5 2	M30 M30 X 2
4.5	0.75	M4.5	33	3.5 2	M33 M33 X 2
5	0.8	M5	36	4 3	M36 M36 X 3
6	1	M6	39	4 3	M39 M39 X 3
7	1	M7			
8	1.25 1	M8 M8 X 1			
10	1.5 1.25	M10 M10 X 1.25			
12	1.75 1.25	M12 M12 X 1.25			

* USA practice is to include the pitch symbol even for the coarse pitch series. Basic designations shown are as specified in ISO Recommendations.

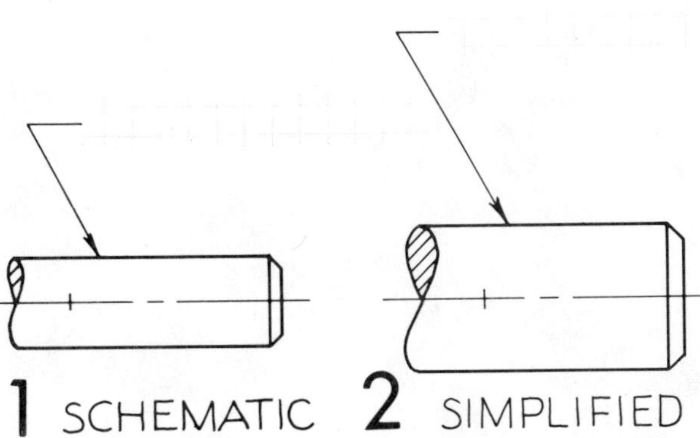

1 SCHEMATIC

2 SIMPLIFIED

3

Nominal Size Diam. (mm)	Pitch P (mm)	Basic Thread Designation	External Thread (Bolt) Tol Class	Allowance	Major Diameter Max	Major Diameter Min	Pitch Diameter Max	Pitch Diameter Min	Pitch Diameter Tol	Minor Diameter Max	Minor Diameter Min b	Internal Thread (Nut) Tol Class	Minor Diameter Min	Minor Diameter Max	Pitch Diameter Min	Pitch Diameter Max	Pitch Diameter Tol	Major Diam. Min
1.6	0.35	M1.6	6g	0.0008	0.0622	0.0589	0.0533	0.0509	0.0024	0.0453	0.0419	6H	0.0481	0.0520	0.0541	0.0574	0.0033	0.0630
1.8	0.35	M1.8	6g	0.0008	0.0701	0.0668	0.0611	0.0588	0.0023	0.0531	0.0498	6H	0.0560	0.0598	0.0620	0.0652	0.0032	0.0709
2	0.4	M2	6g	0.0009	0.0779	0.0743	0.0677	0.0652	0.0025	0.0586	0.0549	6H	0.0617	0.0661	0.0686	0.0720	0.0034	0.0788
2.2	0.45	M2.2	6g	0.0009	0.0858	0.0819	0.0743	0.0716	0.0027	0.0640	0.0601	6H	0.0675	0.0723	0.0752	0.0788	0.0033	[illegible]
12	1.75	M12	6g	0.0014	[illegible]	0.4607	0.4263	0.4205	[illegible]	[illegible]	0.3758	[illegible]	[illegible]	0.4110	0.4277	0.4355	0.0078	0.4725
	1.25	M12 x 1.25	6g	0.0012	0.4713	0.4630	0.4393	0.4342	0.0051	0.4109	0.4023	6H	0.4192	0.4295	0.4405	0.4475	0.0070	[illegible]
14	2	M14	6g	0.0016	0.5496	0.5387	0.4985	0.4923	0.0062	0.4530	0.4412	6H	0.4660	0.4807	0.5001	0.5083	0.0082	0.5512
	1.5	M14 x 1.5	6g	0.0013	0.5499	0.5407	0.5115	0.5061	0.0054	0.4774	0.4677	6H	0.4873	0.4990	0.5129	0.5203	0.0074	0.5512
16	2	M16	6g	0.0016	0.6284	0.6175	0.5772	0.5710	0.0062	0.5318	0.5199	6H	0.5447	0.5594	0.5788	0.5871	0.0083	0.6300
	1.5	M16 x 1.5	6g	0.0014	0.6286	0.6194	0.5903	0.5849	0.0054	0.5561	0.5465	6H	0.5660	0.5777	0.5916	0.5990	0.0074	0.6300
18	2.5	M18	6g	0.0017	0.7070	0.6939	0.6430	0.6364	0.0066	0.5862	0.5725	6H	0.6022	0.6198	0.6448	0.6535	0.0087	0.7087
	1.5	M18 x 1.5	6g	0.0013	0.7074	0.6982	0.6690	0.6636	0.0054	0.6349	0.6252	6H	0.6448	0.6565	0.6704	0.6777	0.0073	0.7087

Tolerance Positions: ISO has established "amounts of allowance" by a series of tolerance position symbols as follows:

External Threads (Bolts):
- Small "e" = large allowance
- Small "g" = small allowance
- Small "h" = no allowance

Internal Threads (Nuts):
- Large "G" = small allowance
- Large "H" = no allowance

The above symbols are used after the Tolerance Grade such as 6g, which designates a "Medium Tolerance Grade" with small allowance for an external thread.

Tolerance Classes: ISO Tolerance classes of fit are determined by selecting one of the three qualities: Fine, Medium or Coarse, combined with one of the three lengths of engagement: Short (S), Normal (N) or Long (L) and applying the proper allowance.

Complete Designation: A complete designation for an ISO Metric Screw Thread comprises in addition to the basic designation, an indentification for the tolerance class. The tolerance class designation is separated from the basic designation by a dash and includes the symbol for the pitch diameter tolerance followed immediately by the symbol for crest diameter tolerance. Each of these symbols shall in turn consist of first a numeral indicating the tolerance grade followed by a letter indicating the tolerance position.

Example:

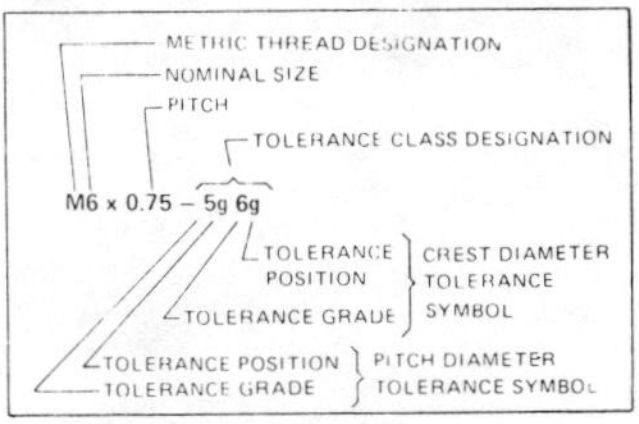

Where the pitch and crest diameter tolerance symbols are identical, the symbol need only be given once and not repeated.

Example:

M20 x 2–6H — PITCH AND CREST DIAMETER TOLERANCE SYMBOL (TOLERANCE EQUAL) — TOLERANCE CLASS DESIGNATION

Length of Engagement Designations: Where considered necessary, the length of engagement group symbol may be added to the tolerance class designation.

Example:

M6–7g 6gL — LENGTH OF ENGAGEMENT GROUP SYMBOL

Designations For Thread Fits: A desired fit between mating threads is indicated by the internal thread class designation followed by the external thread tolerance class designation, separated by a slash.

Examples:

M 6–6H/6g
M20 x 2–6H/5g 6g

4 LETTER A COMPLETE METRIC THREAD NOTE THAT DESIGNATES A 1.8 mm DIA EXTERNAL THREAD. USE THE TABLE ABOVE.

5 LETTER A COMPLETE METRIC THREAD NOTE THAT DESIGNATES A 2 mm DIA INTERNAL THREAD. USE THE TABLE ABOVE.

6 LETTER AND COMPLETE THE METRIC THREAD NOTE THAT DESIGNATES AN 18 mm DIA, COARSE, INTERNAL THREAD WITH A NORMAL (N) LENGTH OF ENGAGEMENT. USE THE TABLE.

7 LETTER A COMPLETE METRIC THREAD NOTE THAT DESIGNATES A 12 mm DIA, COARSE, EXTERNAL THREAD AND NOTE THE FIT BETWEEN TWO MATING THREADS.

JHE

JHE

SKETCH THE DIMENSIONS IN THE VIEWS WHERE THE FEATURES ARE THE MOST DESCRIPTIVE. FOLLOW INSTRUCTIONS A OR B AS ASSIGNED. USE THE 1/8" OR 3mm GRID TO DETERMINE DIMENSIONS.

BASIC DIMENSIONING

A. DIMENSION COMPLETELY OMITTING NUMERALS.
B. DIMENSION COMPLETELY WITH NUMERALS IN DECIMAL FORM.

1 GAUGE

2 SLIDE

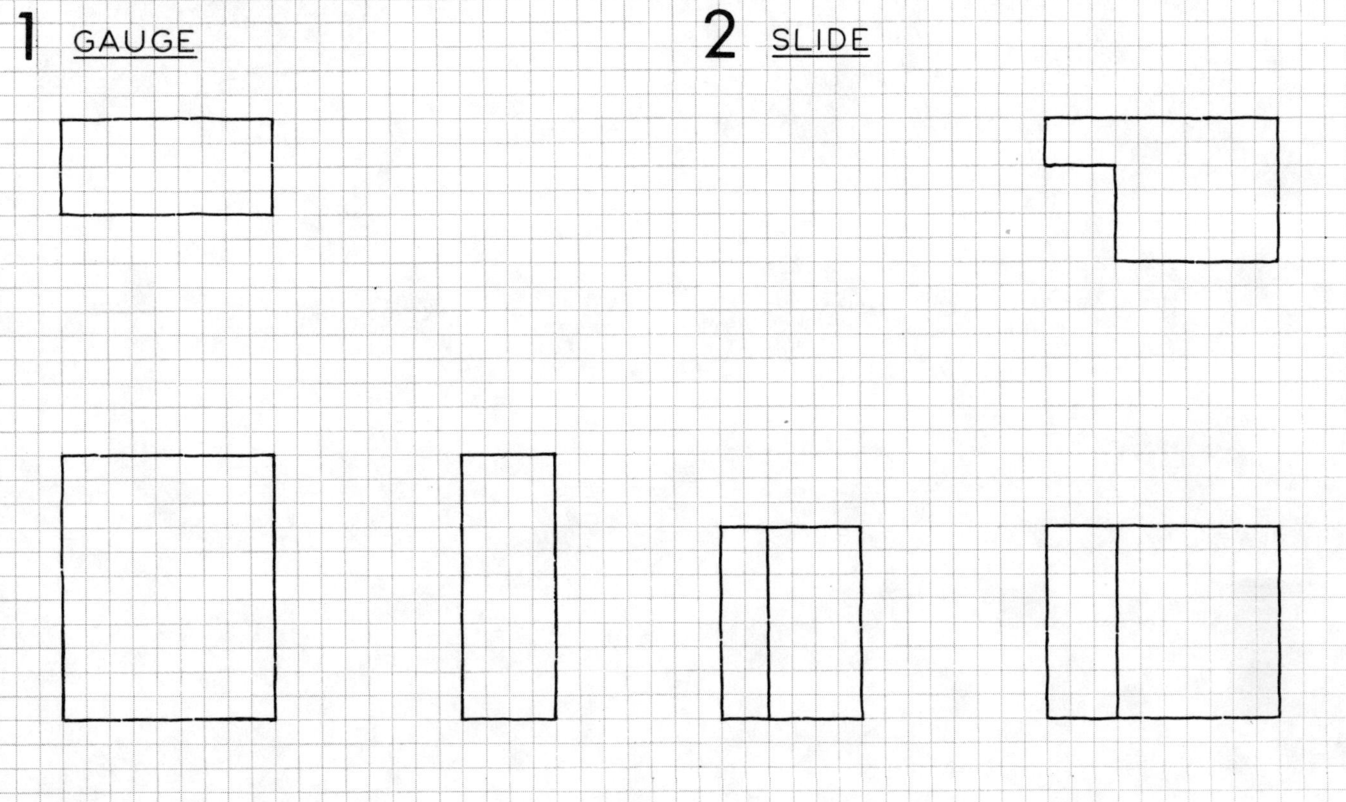

3 WEDGE BLOCK

4 JIG

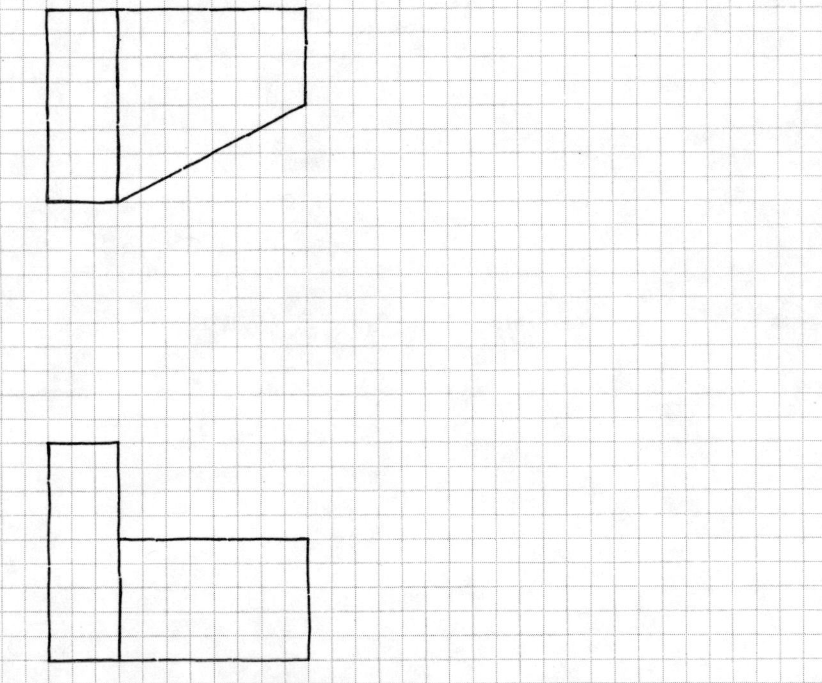

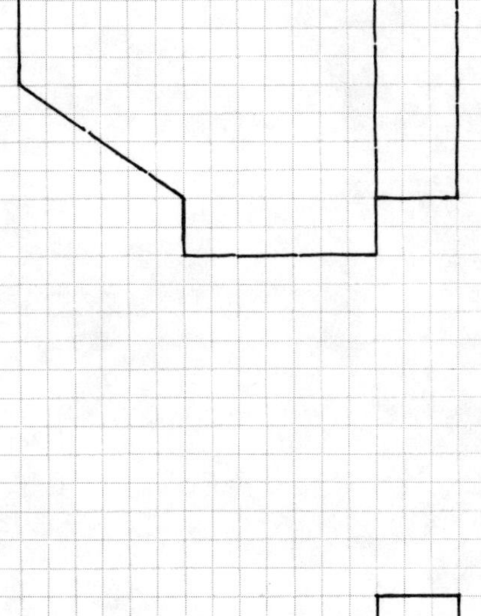

TJK

Graphics & Geometry ©	NAME FILE SEC DATE	MIN.	GRADE	103

BASIC DIMENSIONING

FOLLOW INSTRUCTIONS A OR B, AS ASSIGNED, TO SKETCH THE DIMENSIONS OF THE OBJECTS BELOW. USE THE 1/8" OR 3mm GRID TO DETERMINE DIMENSIONS.

A. DIMENSION COMPLETELY OMITTING NUMERALS.
B. DIMENSION COMPLETELY WITH NUMERALS IN DECIMAL FORM.

1 DISC

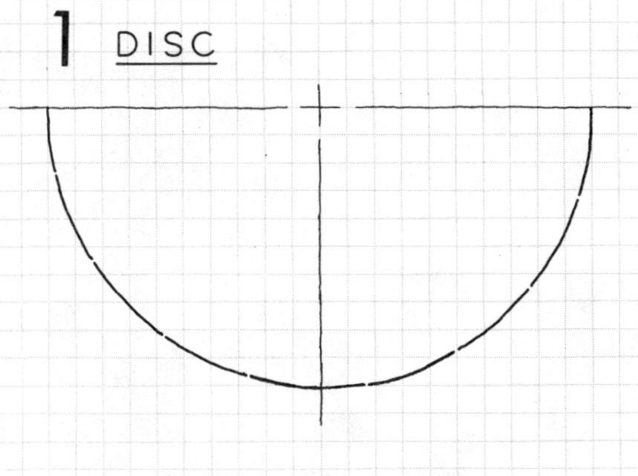

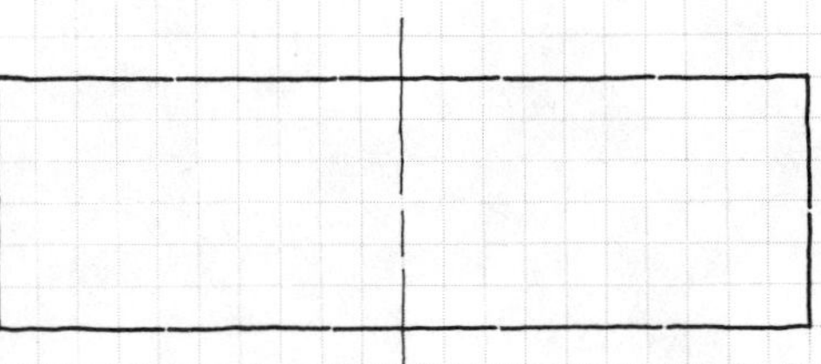

2 PULLEY

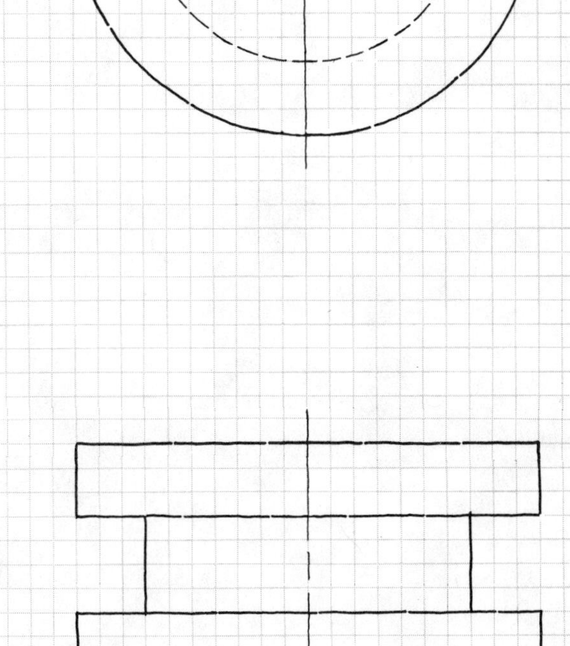

3 SPINDLE

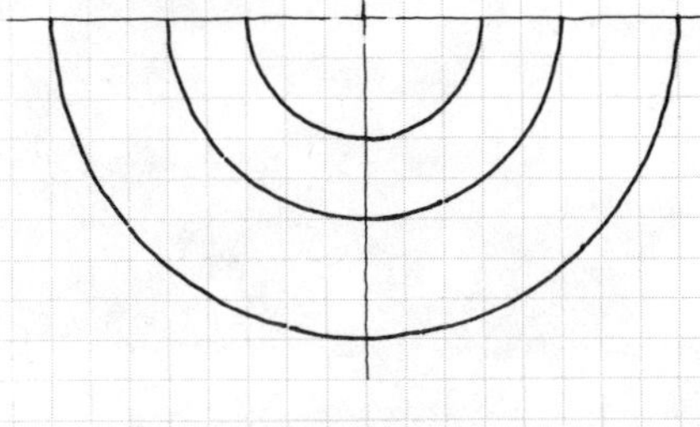

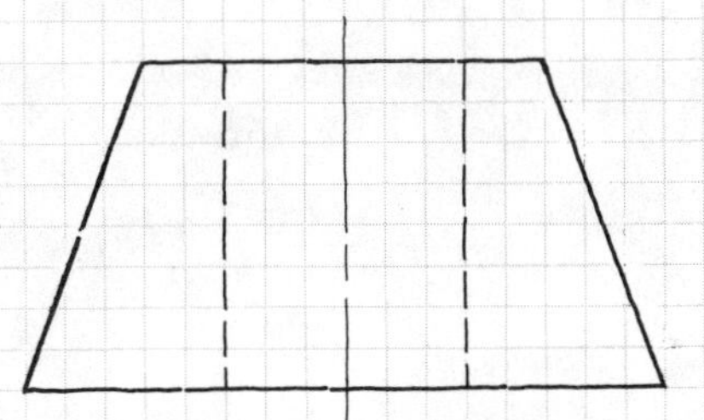

4 SOCKET

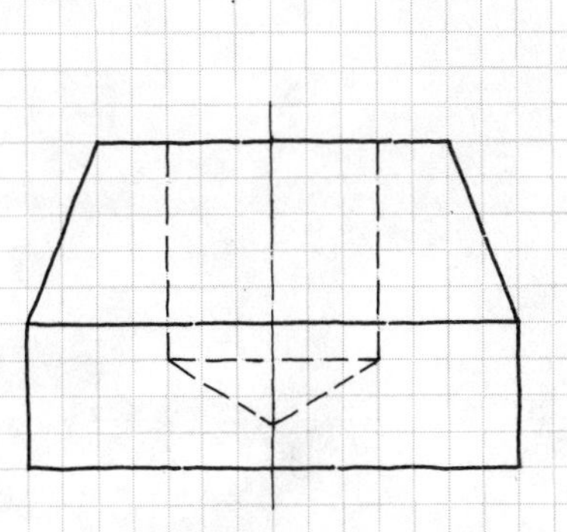

TJK

BASIC DIMENSIONING

1 DIMENSION THE OBJECT BELOW. SCALE: FULL.

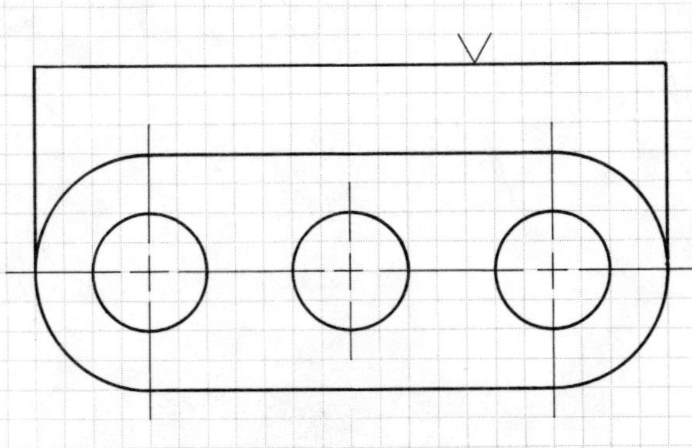

2 DIMENSION THE KEYWAY. SCALE: FULL.

3 USE NOTES TO DIMENSION THE 45° CHAMFER AND THE 96 DP DIAMOND KNURL. SCALE: FULL.

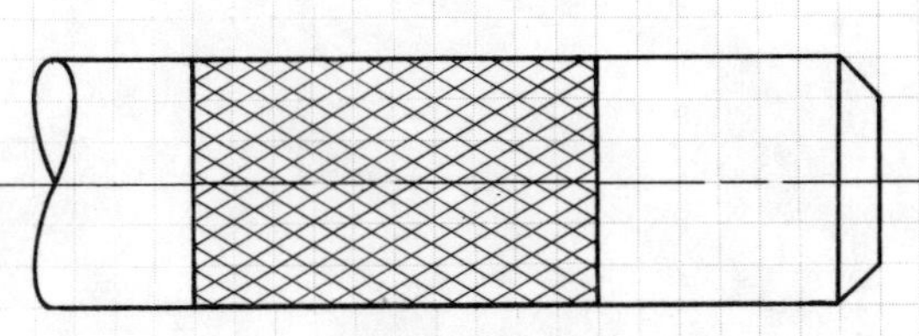

4 THE THROUGH HOLE IS COUNTERBORED .50 INCHES OR 12mm DEEP. DIMENSION BY A NOTE. SCALE: FULL.

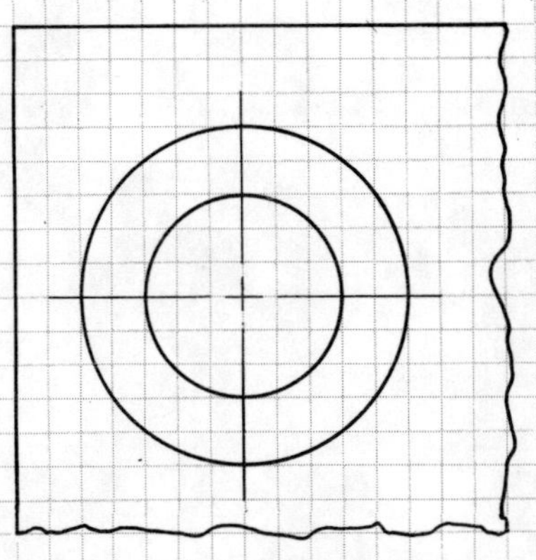

5 USE A NOTE TO DIMENSION THE 82° COUNTERSUNK HOLE. SCALE: FULL.

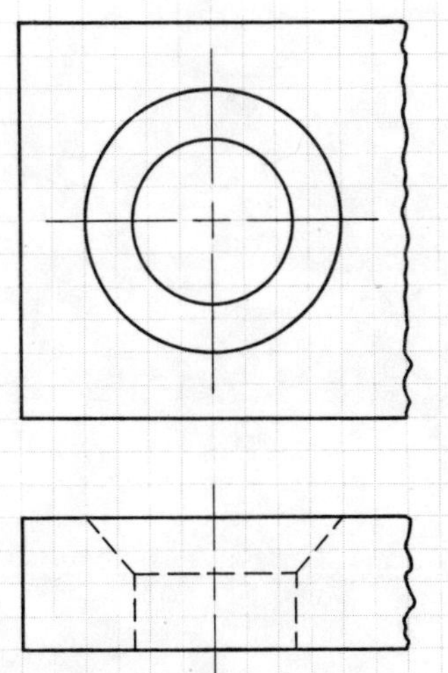

6 USE 4 POINTS TO DIMENSION THE IRREGULAR CURVE. USE THE END LINE AND CENTERLINE FOR DATUM LINES. SCALE: FULL.

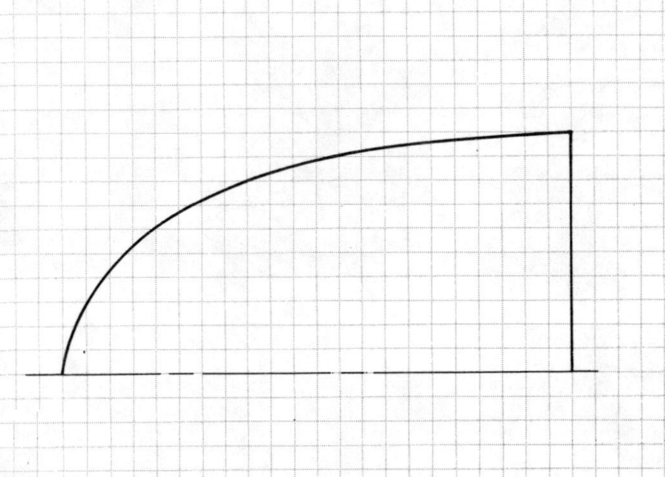

BASIC DIMENSIONING

SKETCH THE COMPLETE DIMENSIONS FOR THE FULL SIZE OBJECTS. FOLLOW INSTRUCTIONS A OR B AS ASSIGNED.

A. DIMENSION COMPLETELY OMITTING NUMERALS.
B. DIMENSION COMPLETELY WITH NUMERALS IN DECIMAL FORM.

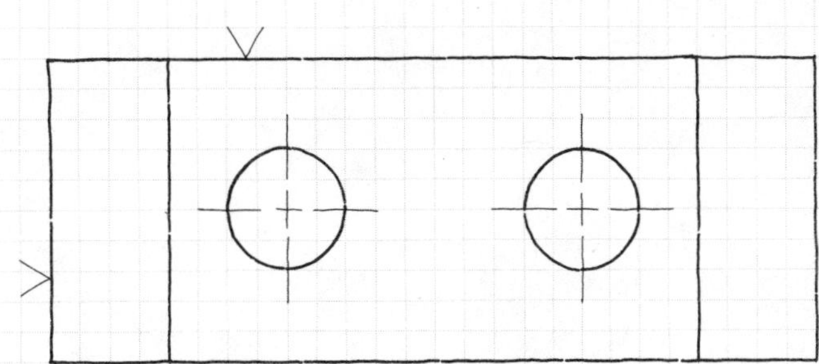

1 GUIDE BLOCK

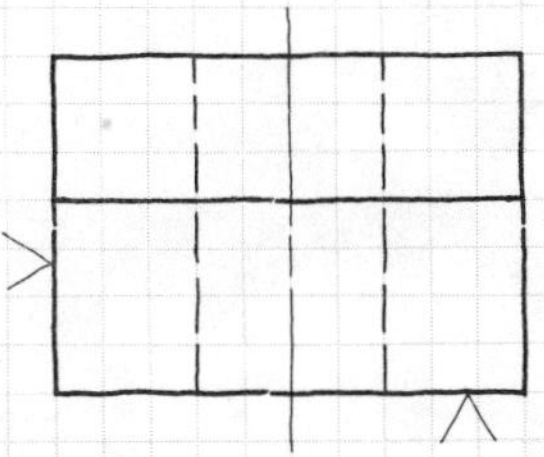

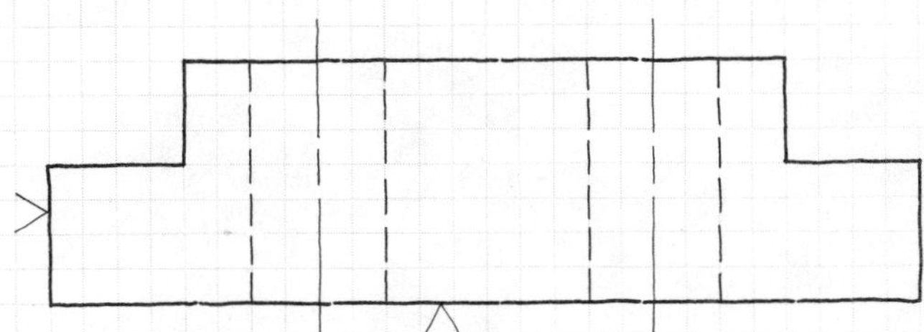

2 CONNECTOR

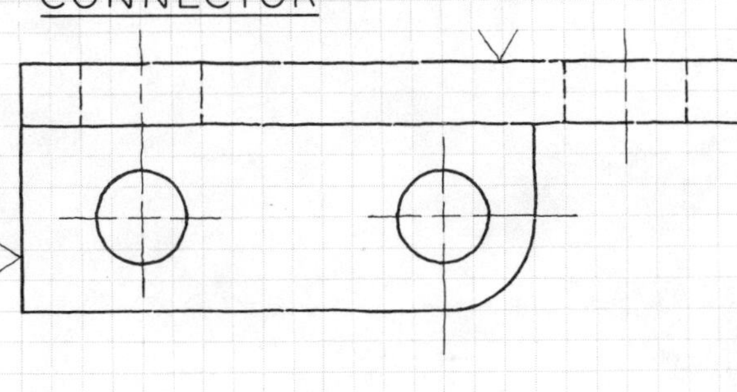

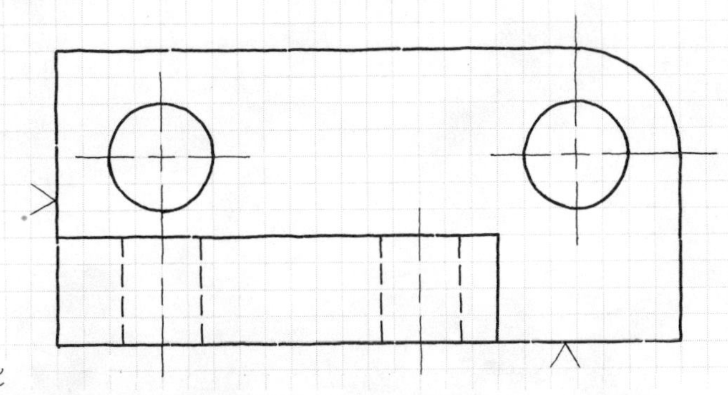

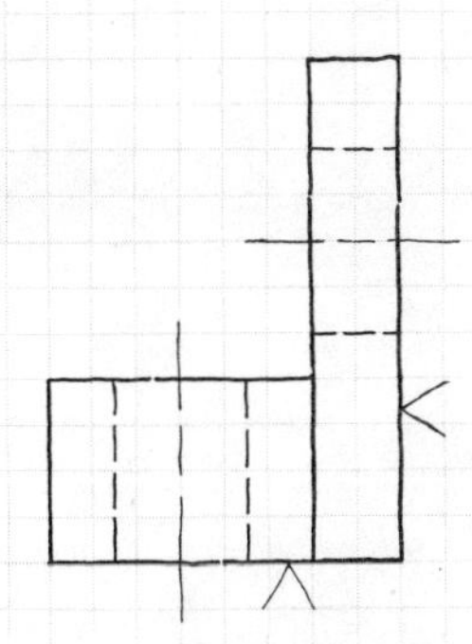

TJK

1 DIMENSION THE FULL SIZE PARTS TO THE NEAREST 0.05 DECIMAL INCH OR WHOLE MILLIMETER.

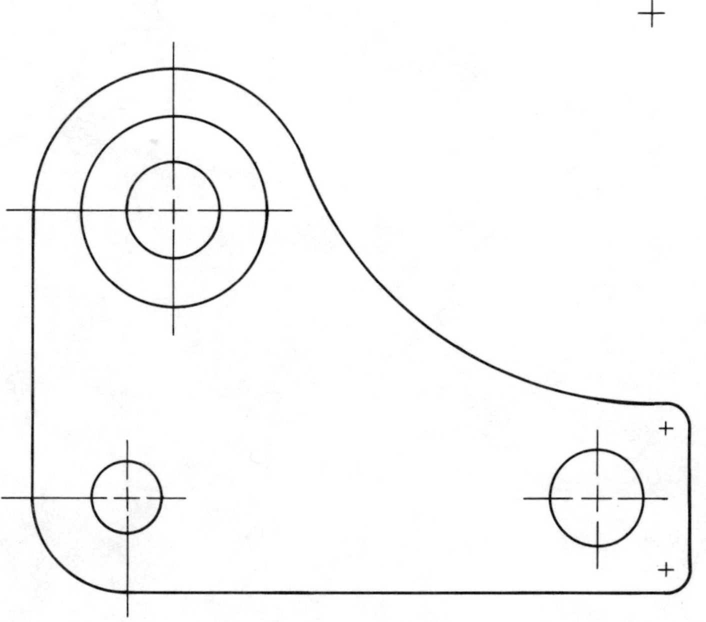

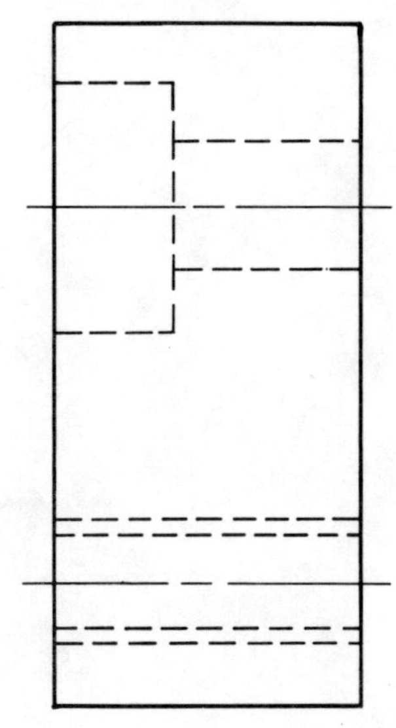

2

3

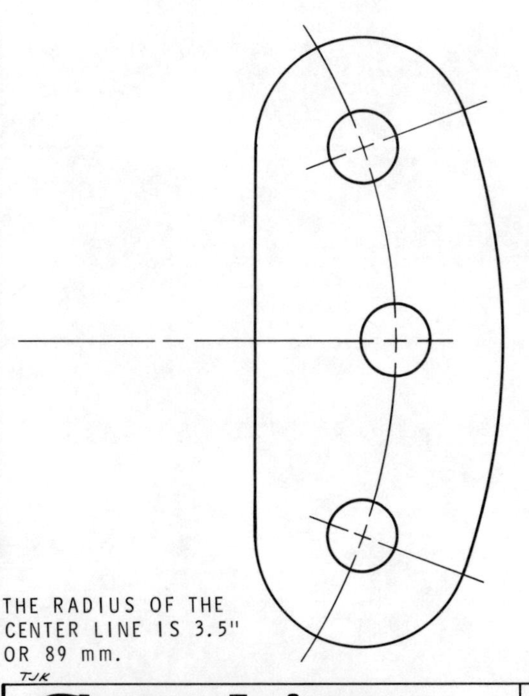

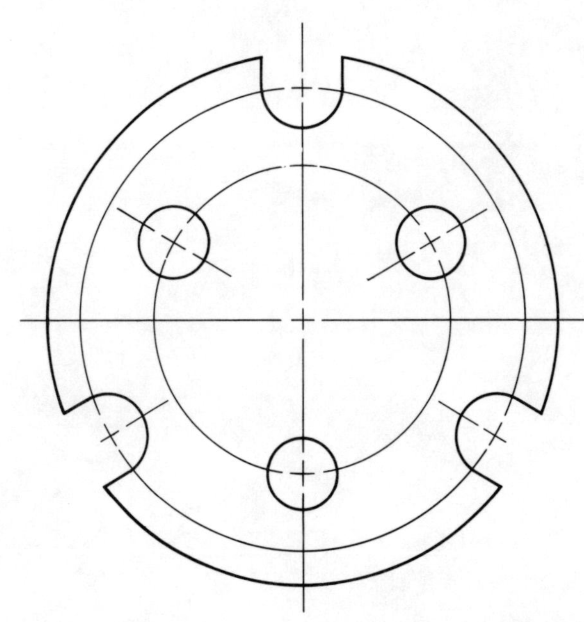

THE RADIUS OF THE CENTER LINE IS 3.5" OR 89 mm.

TJK

Graphics & Geometry ©	NAME FILE SEC DATE	MIN.	GRADE	107

CLASS OF FIT	RC 1 (.6250)	LT 3 (1.0000)	FN 5 (1.5000)
LIMITS OF CLEARANCE OR INTERFERENCE			
SHAFT TOLERANCE			
HOLE TOLERANCE			
ALLOWANCE AT MMC			

TOLERANCE DIMENSION

1 APPLY THE UPPER AND LOWER LIMITS TO THE GIVEN SHAFTS AND HOLES USING THE SPECIFIED AMERICAN STANDARD FITS RC1, LT 3, AND FN 5 RESPECTIVELY.

COMPLETE THE TABLE AT THE LEFT, EXPRESSING ALL DIMENSIONS IN THOUSANDTHS OF AN INCH.

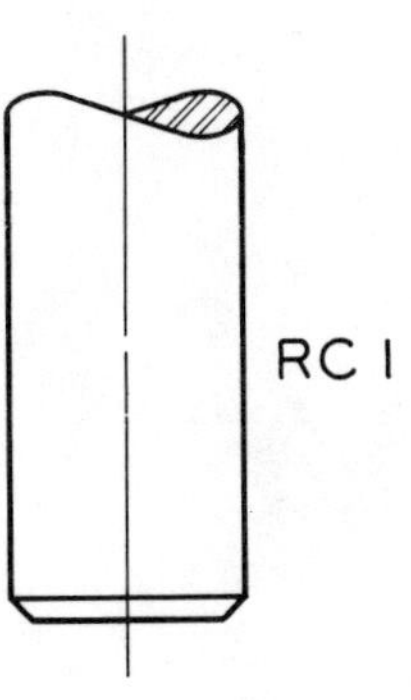

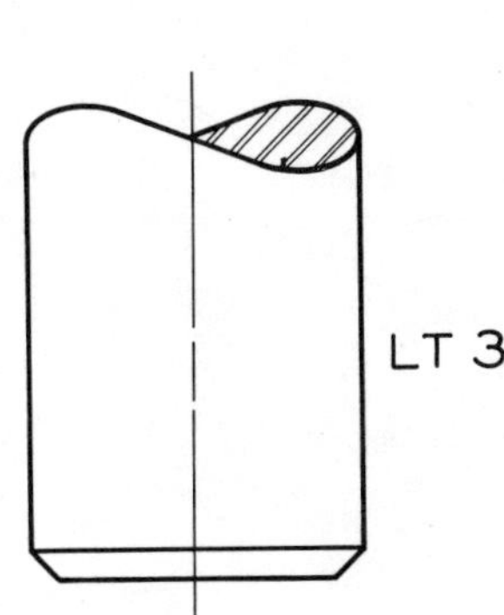

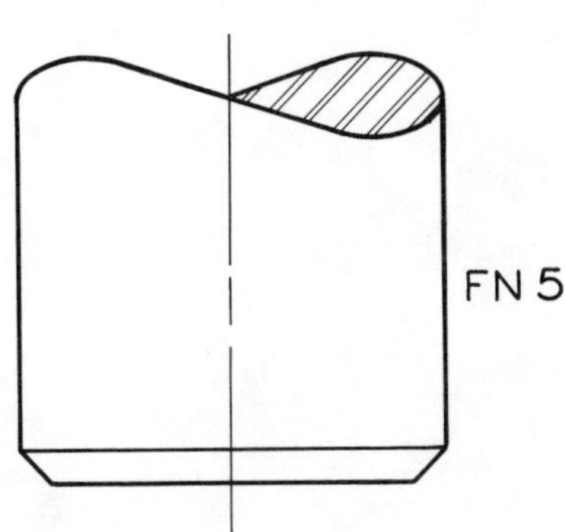

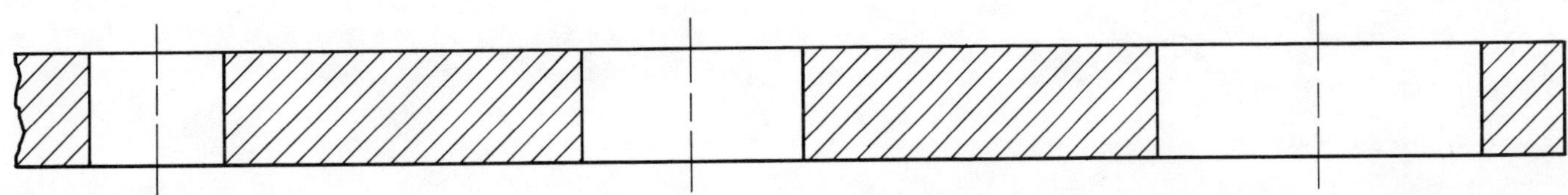

2 LOCATE EACH VERTICAL SURFACE FROM DATUM PLANE A WITH LIMITS OF .000" AND .002".

3 BY USE OF A SURFACE TEXTURE SYMBOL SPECIFY A ROUGHNESS HEIGHT RANGE FOR SURFACE A THAT IS FINISHED BY GRINDING. ALSO SPECIFY A MAXIMUM WAVINESS HEIGHT RATING OF .002" AND WIDTH RATING OF 2 WITH A PERPENDICULAR LAY SYMBOL.

SURFACE B IS TO BE POLISHED AND WILL HAVE A MAXIMUM WAVINESS HEIGHT OF .002" AND A WIDTH RATING OF 2" WITH THE LAY PERPENDICULAR TO THE FRONT SURFACE.AND WITH A ROUGHNESS—WIDTH CUTOFF RATING OF .100.

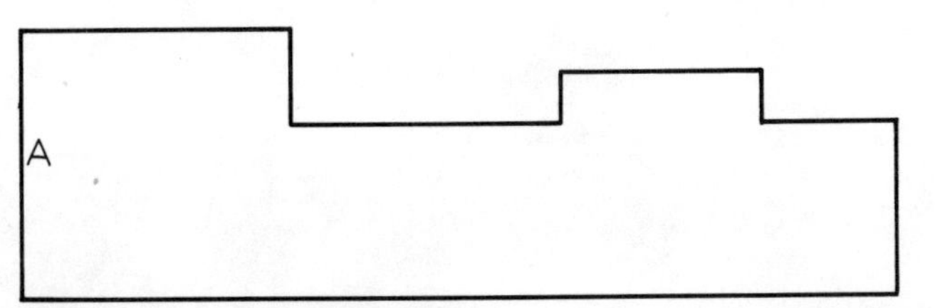

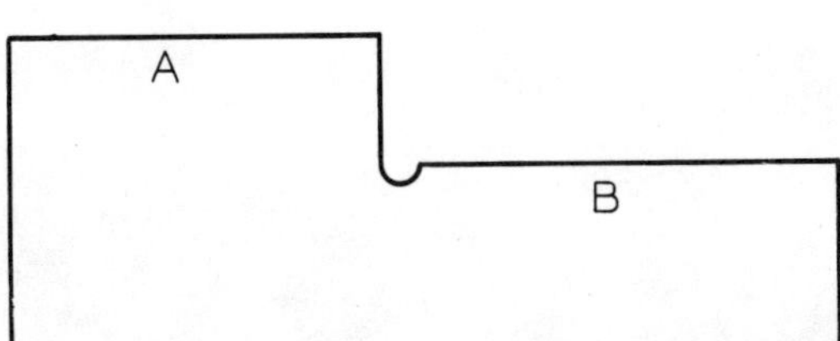

TOLERANCES OF FORM

1 USING SYMBOLS, DIMENSION SURFACE B TO HAVE AN ANGULARITY TOLERANCE OF .02 (0.51mm) USING SURFACE A AS THE DATUM PLANE.

2 USING SYMBOLS, DIMENSION SURFACES C AND D TO BE PARALLEL TO DATUM E WITHIN .008 (0.20 mm).

3 USING SYMBOLS, DIMENSION SURFACES F AND G TO BE PERPENDICULAR TO DATUM H WITHIN .04 (1.02 mm).

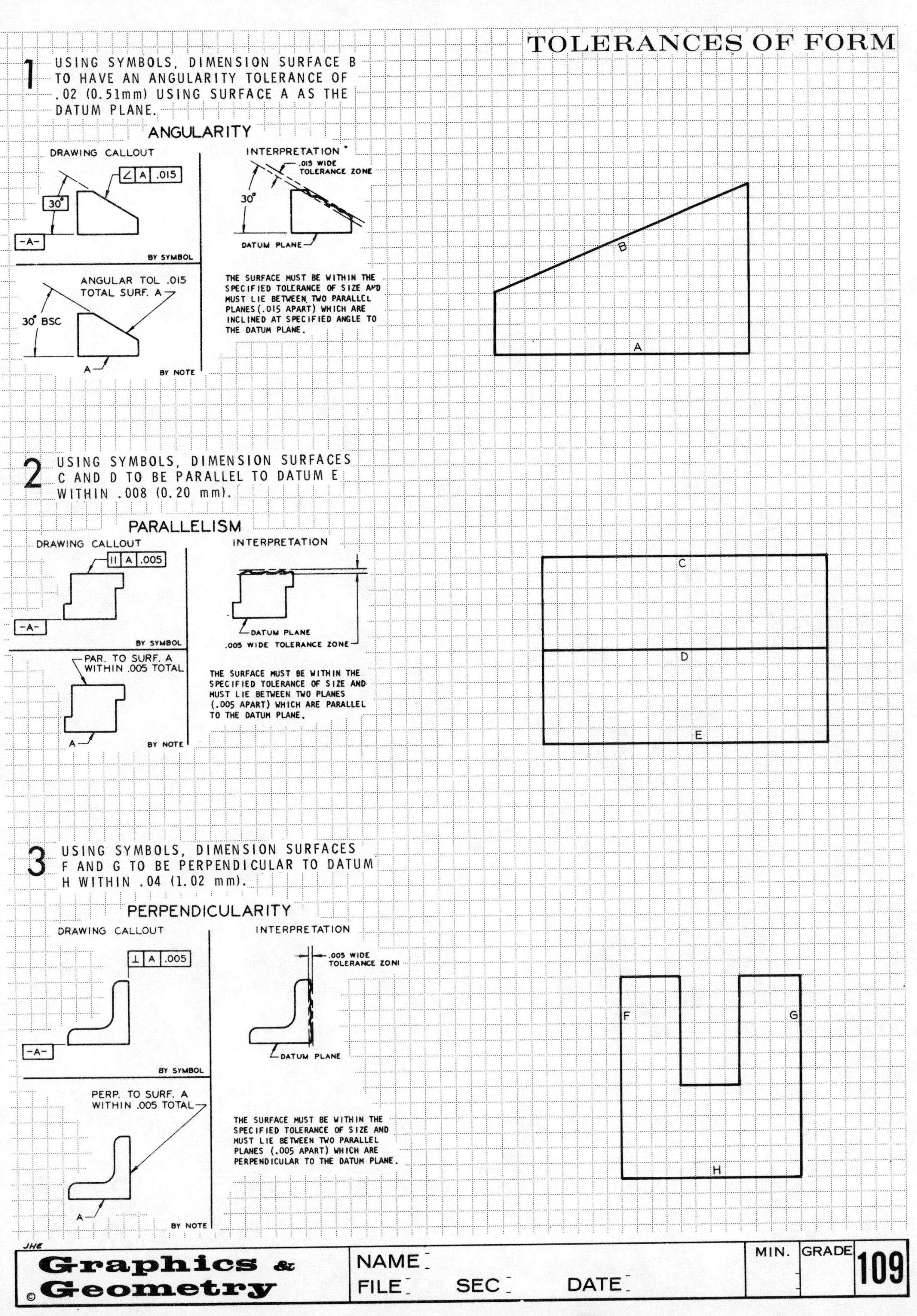

JHE

Graphics & Geometry	NAME FILE SEC DATE	MIN.	GRADE	109

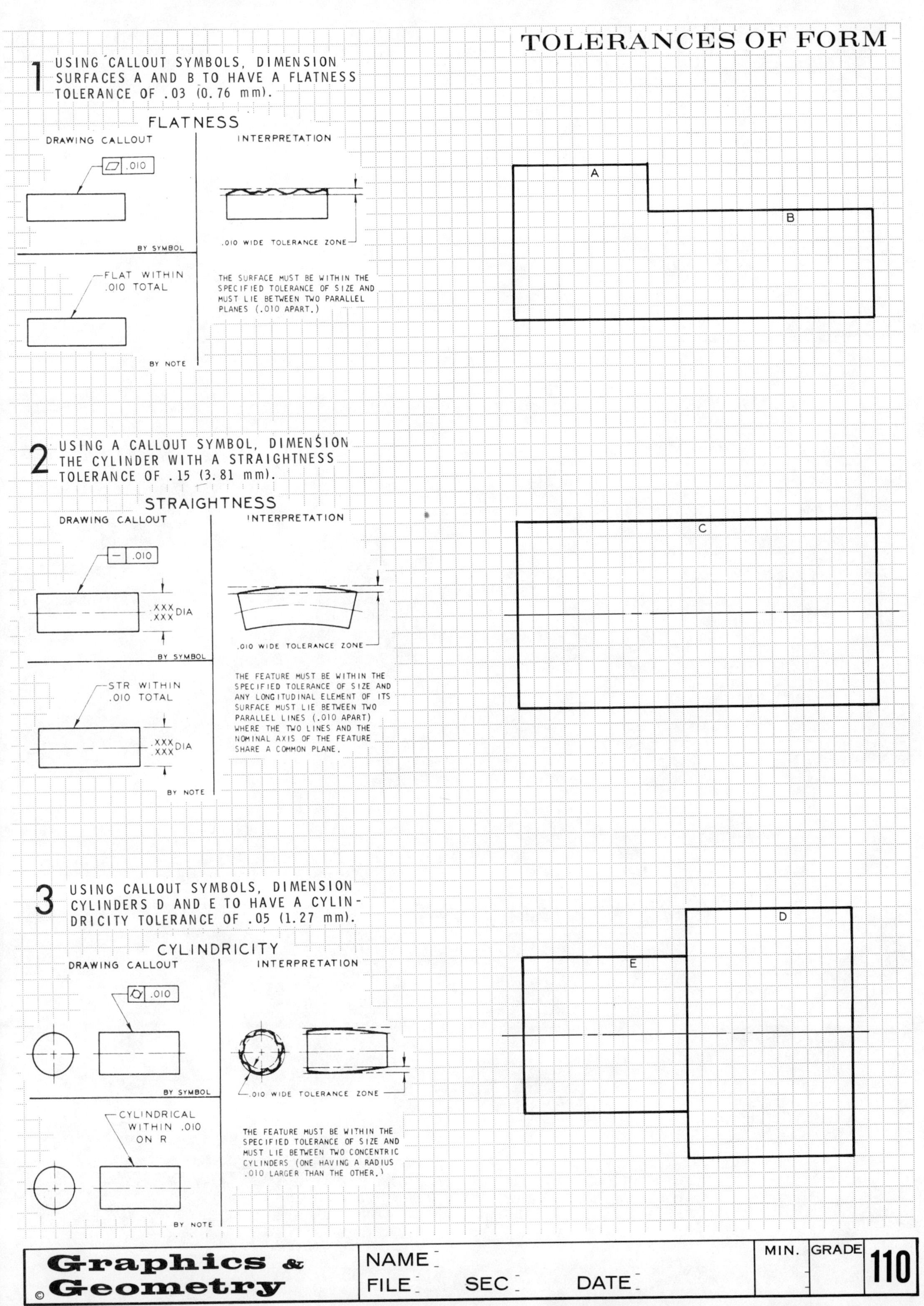
TOLERANCES OF FORM
1 USING CALLOUT SYMBOLS, DIMENSION SURFACES A AND B TO HAVE A FLATNESS TOLERANCE OF .03 (0.76 mm).
FLATNESS
DRAWING CALLOUT
INTERPRETATION
.010
BY SYMBOL
.010 WIDE TOLERANCE ZONE
FLAT WITHIN .010 TOTAL
THE SURFACE MUST BE WITHIN THE SPECIFIED TOLERANCE OF SIZE AND MUST LIE BETWEEN TWO PARALLEL PLANES (.010 APART.)
BY NOTE
A
B
2 USING A CALLOUT SYMBOL, DIMENSION THE CYLINDER WITH A STRAIGHTNESS TOLERANCE OF .15 (3.81 mm).
STRAIGHTNESS
DRAWING CALLOUT
INTERPRETATION
.010
.XXX .XXX DIA
BY SYMBOL
.010 WIDE TOLERANCE ZONE
STR WITHIN .010 TOTAL
THE FEATURE MUST BE WITHIN THE SPECIFIED TOLERANCE OF SIZE AND ANY LONGITUDINAL ELEMENT OF ITS SURFACE MUST LIE BETWEEN TWO PARALLEL LINES (.010 APART) WHERE THE TWO LINES AND THE NOMINAL AXIS OF THE FEATURE SHARE A COMMON PLANE.
.XXX .XXX DIA
BY NOTE
C
3 USING CALLOUT SYMBOLS, DIMENSION CYLINDERS D AND E TO HAVE A CYLINDRICITY TOLERANCE OF .05 (1.27 mm).
CYLINDRICITY
DRAWING CALLOUT
INTERPRETATION
.010
BY SYMBOL
.010 WIDE TOLERANCE ZONE
CYLINDRICAL WITHIN .010 ON R
THE FEATURE MUST BE WITHIN THE SPECIFIED TOLERANCE OF SIZE AND MUST LIE BETWEEN TWO CONCENTRIC CYLINDERS (ONE HAVING A RADIUS .010 LARGER THAN THE OTHER.)
BY NOTE
D
E
Graphics & Geometry
©
NAME
FILE
SEC
DATE
MIN.
GRADE
110

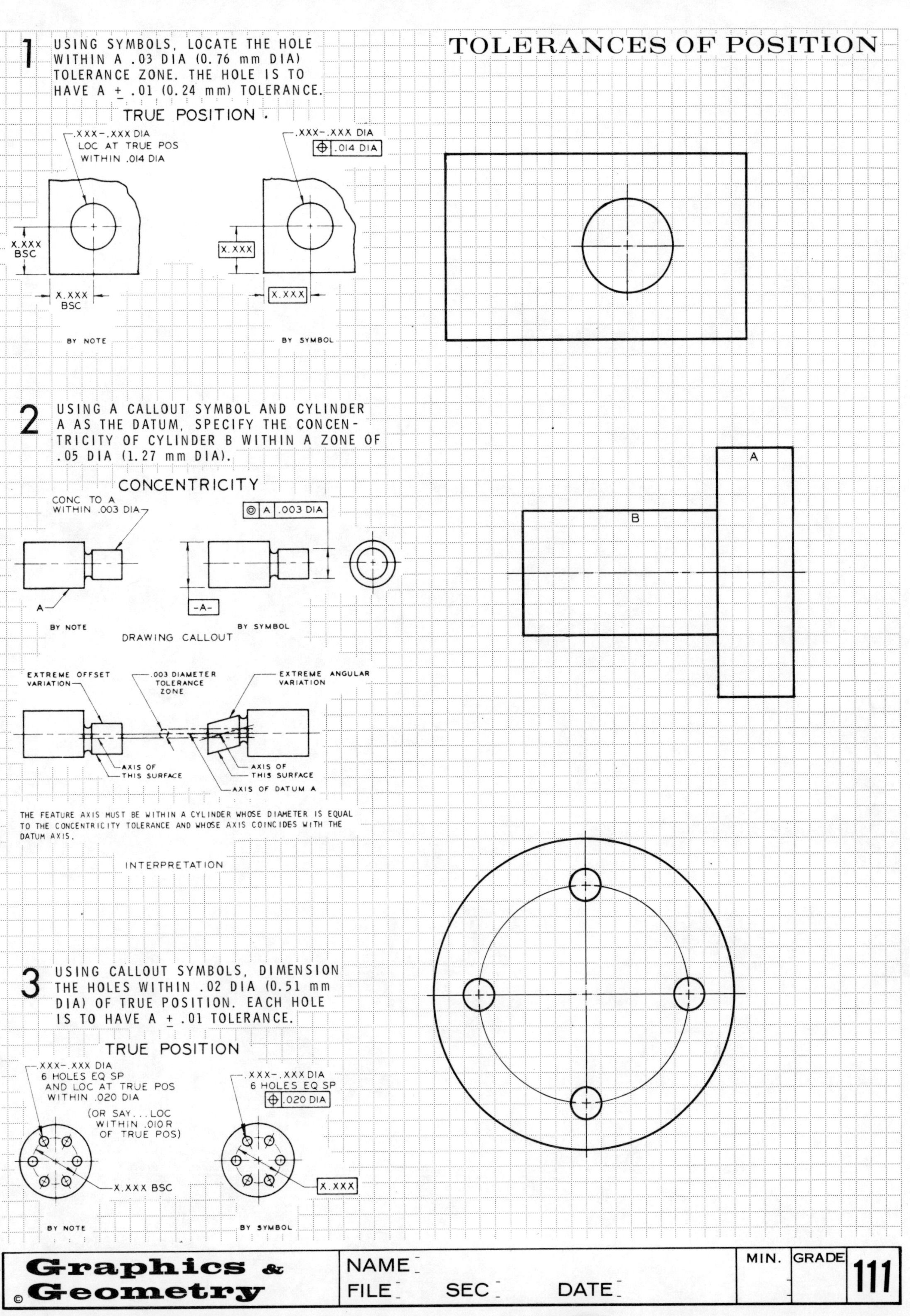
TOLERANCES OF POSITION
1 USING SYMBOLS, LOCATE THE HOLE WITHIN A .03 DIA (0.76 mm DIA) TOLERANCE ZONE. THE HOLE IS TO HAVE A ± .01 (0.24 mm) TOLERANCE.
TRUE POSITION
.XXX-.XXX DIA
LOC AT TRUE POS
WITHIN .014 DIA
.XXX-.XXX DIA
.014 DIA
X.XXX
BSC
X.XXX
X.XXX
BSC
X.XXX
BY NOTE
BY SYMBOL
2 USING A CALLOUT SYMBOL AND CYLINDER A AS THE DATUM, SPECIFY THE CONCENTRICITY OF CYLINDER B WITHIN A ZONE OF .05 DIA (1.27 mm DIA).
CONCENTRICITY
CONC TO A
WITHIN .003 DIA
A .003 DIA
A
-A-
BY NOTE
BY SYMBOL
DRAWING CALLOUT
A
B
EXTREME OFFSET VARIATION
.003 DIAMETER TOLERANCE ZONE
EXTREME ANGULAR VARIATION
AXIS OF THIS SURFACE
AXIS OF THIS SURFACE
AXIS OF DATUM A
THE FEATURE AXIS MUST BE WITHIN A CYLINDER WHOSE DIAMETER IS EQUAL TO THE CONCENTRICITY TOLERANCE AND WHOSE AXIS COINCIDES WITH THE DATUM AXIS.
INTERPRETATION
3 USING CALLOUT SYMBOLS, DIMENSION THE HOLES WITHIN .02 DIA (0.51 mm DIA) OF TRUE POSITION. EACH HOLE IS TO HAVE A ± .01 TOLERANCE.
TRUE POSITION
.XXX-.XXX DIA
6 HOLES EQ SP
AND LOC AT TRUE POS
WITHIN .020 DIA
(OR SAY...LOC
WITHIN .010 R
OF TRUE POS)
X.XXX BSC
.XXX-.XXX DIA
6 HOLES EQ SP
.020 DIA
X.XXX
BY NOTE
BY SYMBOL
Graphics & Geometry
©
NAME
FILE
SEC
DATE
MIN.
GRADE
111

DIMENSIONING

Completely sketch the dimensions of the universal microstop in a manner that meets the following requirements. The basic PIN size is .1875 (4.6mm) diameter. The limits of the tolerance of the BLOCK and JAW are +.00 and -.01 (+0.0 and -0.3mm). The basic dimension of the BLOCK is .49 (12.44mm). The allowance between the two should be .01 (0.25mm).

PIN and the JAW FN1

PIN and the BLOCK RC9

SHAFT SCREW (sheet 62, basic size .2500) and the BLOCK RC9

Use the limits of the PIN which were established using the FN1 fit and the limits of clearance of the RC9 fit to determine the limits of the hole in the BLOCK.

Save this drawing for use with sheets 118-120.

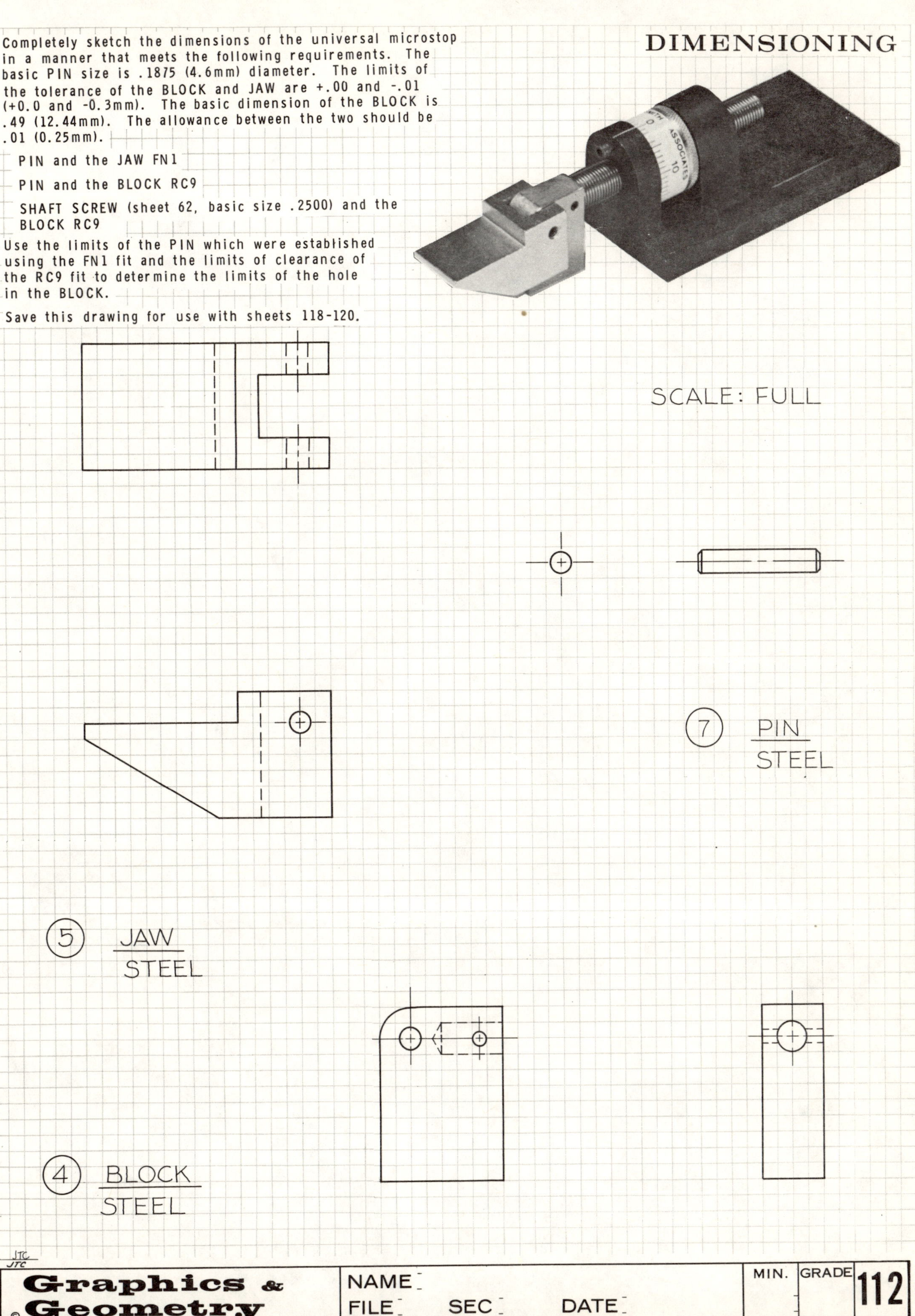

DIMENSIONING

DIMENSION THE BASE OF THE UNIVERSAL MICRO-STOP IN ACCORDANCE WITH THE FOLLOWING SPECIFICATIONS:

1. DETERMINE DIMENSIONS BY MEASURING THE FULL SIZE VIEWS TO THE NEAREST MILLIMETER OR TO TWO PLACE DECIMAL FRACTIONS IN INCHES.
2. THREADED HOLE IN LUG #6-32UNC-2B OR M3.5 X 0.6; THREADED HOLE IN BASE 7/16-14UNC-2B OR M12 X 1.75.
3. INNER SURFACES OF THE LUGS ARE PERPENDICULAR TO SURFACE "A" TO .005 INCH OR 0.15mm.
4. TOLERANCE BETWEEN INSIDE OF LUGS +.01, -.00 INCH OR +0.25, -0.00mm.
5. FIT ON THE LARGE HOLES IN THE LUGS TO BE RC9.
6. FIT ON HOLE AT THE BOTTOM OF THE BASE TO BE RC9.
7. RETAIN THIS LAYOUT FOR USE ON FOLDOUTS.

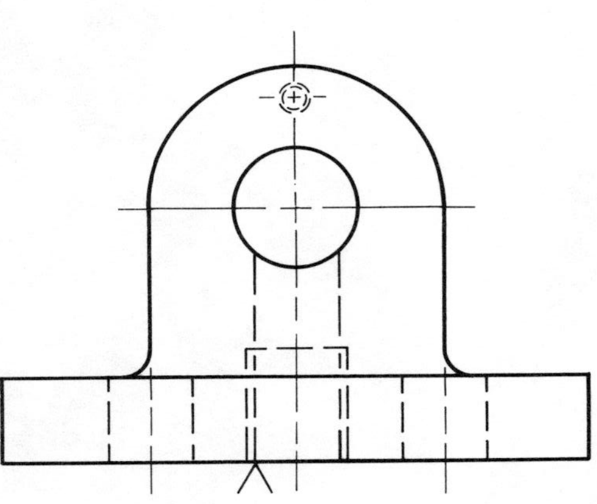

MATL: C.I.

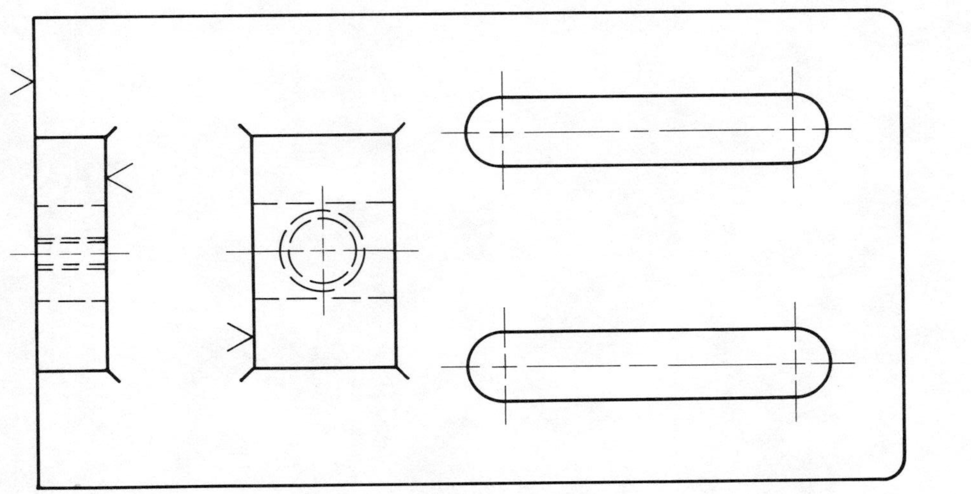

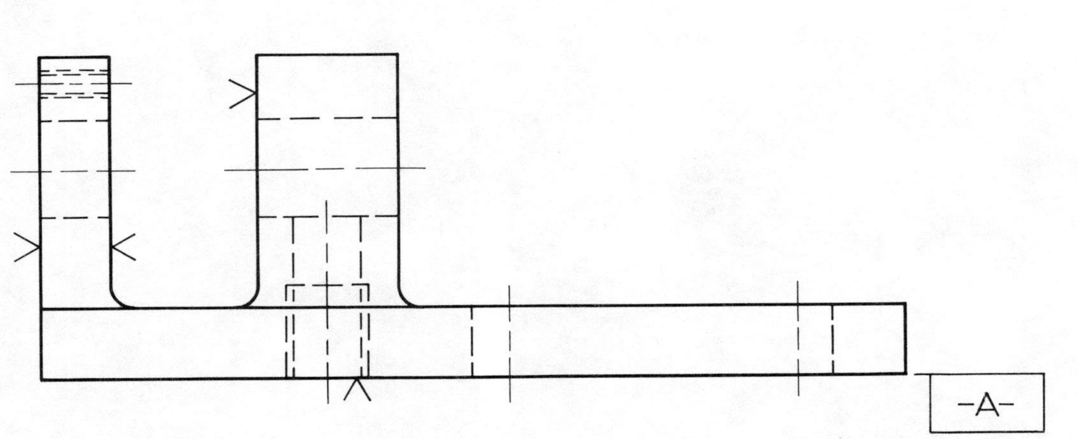

COMPUTER GRAPHICS

1 IN MANY COMPUTER GRAPHIC SYSTEMS THE STATEMENT CALL PLOT (X, Y, IPEN), WILL CAUSE THE PLOTTER TO MOVE TO THE COORDINATE SPECIFIED BY THE X AND Y VALUES. IF IPEN IS EQUAL TO 2, THE PLOTTER PEN WILL BE DOWN AND A LINE WILL BE DRAWN. IF THE IPEN VALUE IS 3, THE PLOTTER WILL MOVE WITH THE PEN IN THE UP POSITION AND A LINE WILL NOT BE DRAWN. USING CALL PLOT STATEMENTS, DRAW THE LETTER A.

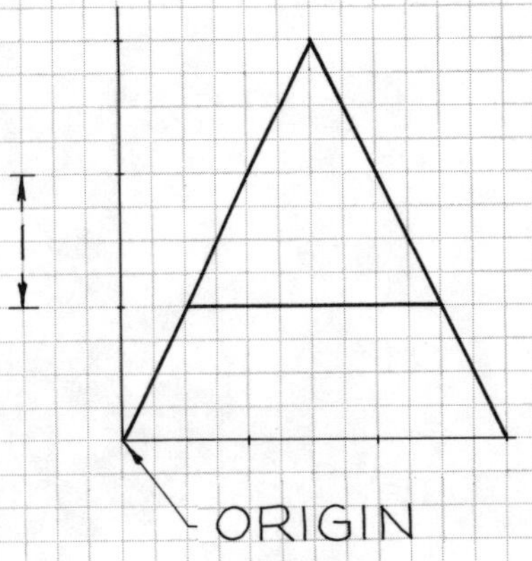

2 WRITE THE CALL PLOT STATEMENTS NECESSARY TO DRAW THE RECTANGLE BELOW.

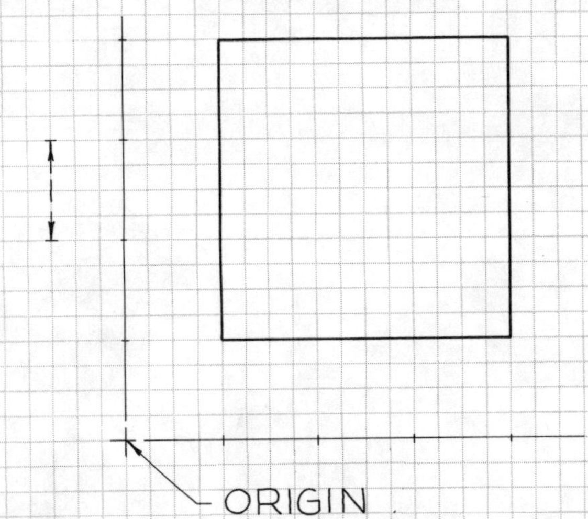

3 WRITE THE CALL PLOT STATEMENTS NECESSARY TO DRAW A RECTANGLE IN TERMS OF THE VARIABLES X, Y, W, AND H. THIS COULD BE USED AS A SUBROUTINE TO DRAW A RECTANGLE OF ANY SIZE ANYWHERE ON THE SHEET.

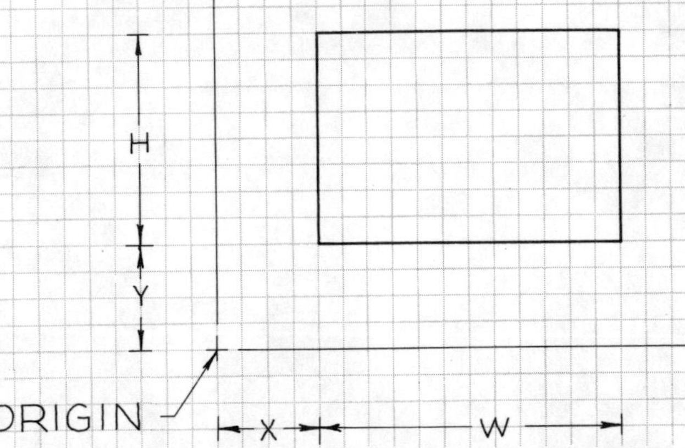

COMPUTER GRAPHICS

THE STEPS LISTED BELOW WILL LEAD TO THE DEVELOPMENT OF A PROGRAM THAT WILL CONSTRUCT BAR GRAPHS. THE USER SHOULD SPECIFY THE HEIGHT AND WIDTH OF THE GRAPH, THE NUMBER OF BARS, THE UPPERMOST SCALE VALUE, AND THE DATA EACH BAR IS TO REPRESENT.

1 USING THE RECTANGLE SUBROUTINE DEVELOPED ON THE PREVIOUS SHEET DRAW THE OUTER RECTANGLE. LOCATE ITS LOWER LEFT CORNER AT COORDINATE (0, 0). ITS HEIGHT SHOULD BE EQUAL TO THE BAR GRAPH HEIGHT, BGH, AND ITS WIDTH SHOULD BE EQUAL TO THE BAR GRAPH WIDTH, BGW.

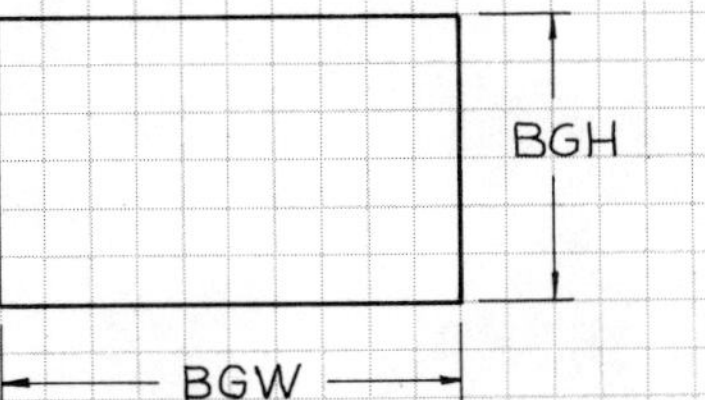

2 ASSUME THAT THE BAR WIDTH, BW, IS TWICE THE SPACE WIDTH, S. THE SPACE WIDTH CAN BE FOUND BY EQUATING THE BAR GRAPH WIDTH AND THREE TIMES THE NUMBER OF BARS, NOBARS, TIMES THE SPACE WIDTH PLUS ONE SPACE AND SOLVING FOR THE SPACE WIDTH. THE BAR WIDTH IS TWICE THE SPACE WIDTH.

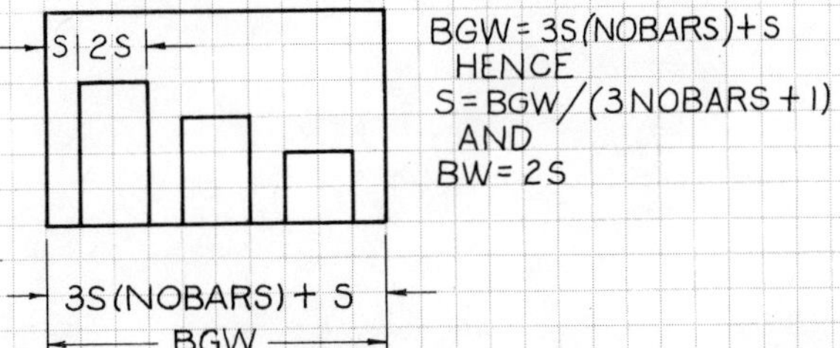

3 CALCULATE A SCALE FACTOR, SF, BY DIVIDING THE BAR GRAPH HEIGHT, BGH, BY THE UPPERMOST SCALE VALUE, TOPSC. THE HEIGHT OF EACH BAR WOULD THEN BE EQUAL TO THE SCALE FACTOR TIMES THE VALUE OF THE DATA. IN THIS PROGRAM THE VALUE OF THE FIRST BAR SHOULD BE READ AND THE BAR DRAWN BEFORE READING THE VALUE OF THE SECOND.

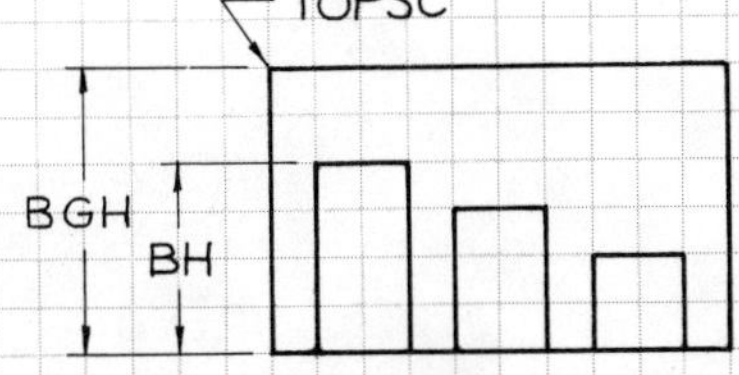

SF = BGH / TOPSC
AND
BH = SF(DATA)

4 THE POSITION OF EACH BAR MAY BE CALCULATED BY OBSERVING THAT THE FIRST BAR IS LOCATED AT COORDINATE (S, 0) AND EACH SUCCEEDING BAR IS LOCATED AT THE PREVIOUS 'X' COORDINATE PLUS THREE TIMES THE SPACE WIDTH.

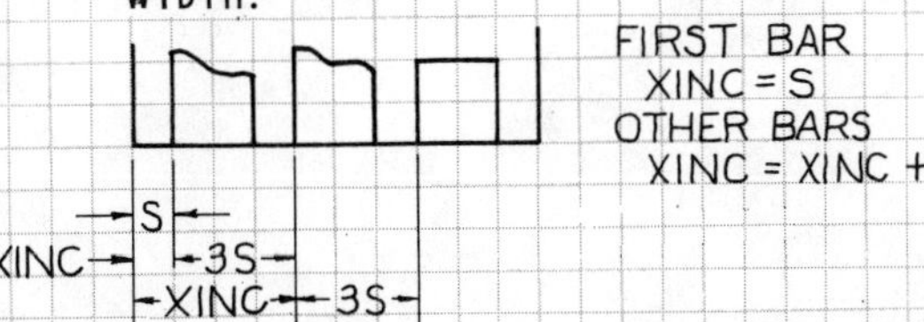

COMPUTER GRAPHICS

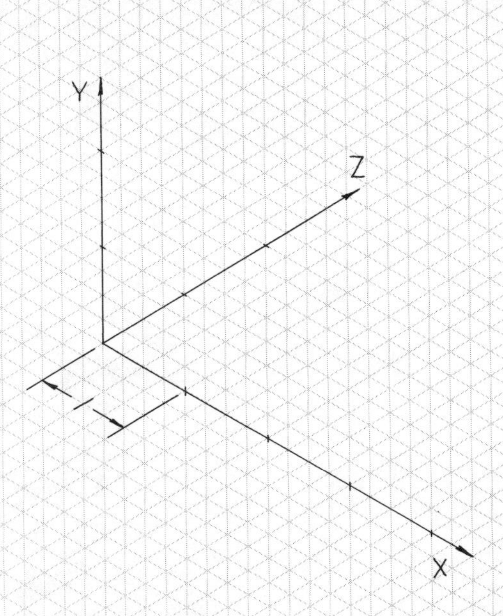

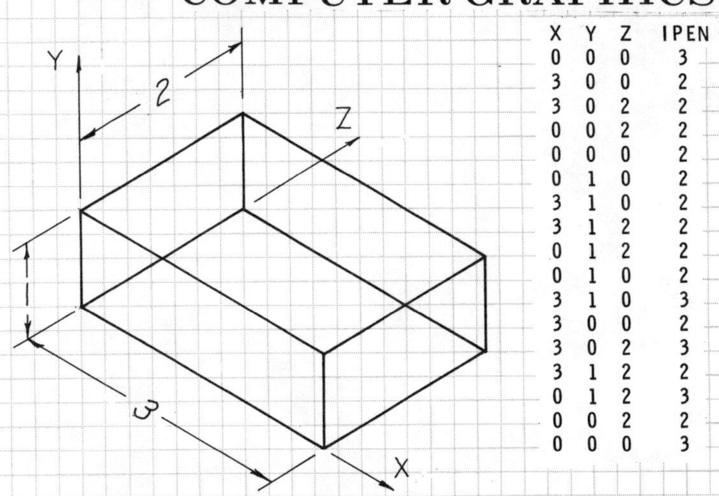

X	Y	Z	IPEN
0	0	0	3
3	0	0	2
3	0	2	2
0	0	2	2
0	0	0	2
0	1	0	2
3	1	0	2
3	1	2	2
0	1	2	2
0	1	0	2
3	1	0	3
3	0	0	2
3	0	2	3
3	1	2	2
0	1	2	3
0	0	2	2
0	0	0	3

1. WHEN PLOTTING THREE DIMENSIONAL OBJECTS THE X-Y-, AND Z- COORDINATES OF THE DESTINATION OF THE PLOTTER AS WELL AS THE IPEN VALUE (2 - DOWN, AND 3 - UP) MUST BE GIVEN. VERIFY THAT THE POINTS AND THE IPEN VALUES ARE CORRECT BY PLOTTING THEM IN AN ISOMETRIC SKETCH. USE THE LEFT HANDED COORDINATE SYSTEM SHOWN

2. IN ORDER TO CONSTRUCT AN OBLIQUE DRAWING OF AN OBJECT THE THREE DIMENSIONAL (X, Y, Z) COORDINATE SYSTEM MUST BE CHANGED TO A TWO DIMENSIONAL COORDINATE SYSTEM (XPLOT, YPLOT) FOR THE PLOTTER. THE ANGLE OF THE RECEDING AXIS AND THE SCALE FACTOR FOR THE MEASUREMENTS ALONG THIS AXIS MUST BE DEFINED. IN THE CASE OF A CAVALIER THE SCALE FACTOR SHOULD BE 1 AND IT SHOULD BE .5 FOR A CABINET. NOTE THAT THE XPLOT AND YPLOT COORDINATES COULD BE EXPRESSED BY THE ALGEBRAIC EQUATIONS BELOW.

WRITE A FORTRAN PROGRAM THAT WILL READ THE DATA IN PROBLEM 1 AND PLOT THE CAVALIER OBLIQUE DRAWING OF THE OBJECT.

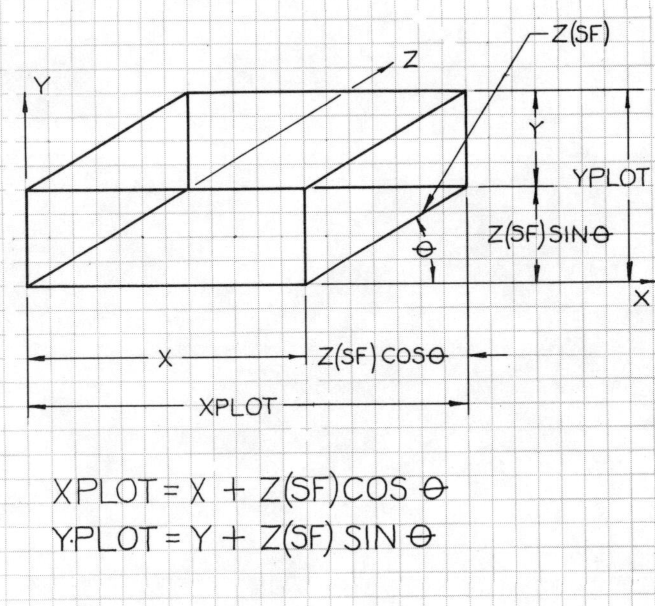

$$XPLOT = X + Z(SF)\cos\theta$$
$$YPLOT = Y + Z(SF)\sin\theta$$

WORKING DRAWINGS ARE USED BY ENGINEERS AND TECHNICIANS TO CONVEY IDEAS TO THOSE WHO ARE TO EXECUTE THEIR DESIGNS. SUCH DRAWINGS SHOULD INCLUDE: ENOUGH GRAPHICAL REPRESENTATIONS TO COMPLETELY DESCRIBE THE SHAPE, SUFFICIENT DIMENSIONS TO SHOW THE SIZE, INDICATIONS OF THE MATERIALS TO BE USED, METHODS OF FINISH, EXPLANATORY NOTES, AN ASSEMBLY AND A PARTS LIST.

ASSUME THAT AN ENGINEER HAS CHECKED YOUR DRAWINGS OF THE UNIVERSAL MICROSTOP THAT WERE SOLVED ON SHEETS 101, 112, AND 113. IN YOUR ROLE AS DRAFTSMAN, MAKE A SET OF WORKING DRAWINGS BY TRACING THE CORRECTED SOLUTIONS ON THE FOLDOUT SHEETS 118, 119, AND 120.

	FIT	BASIC DIA	PART	and	PART
1.	FN 1	0.188" (4.78)	PIN (7)		JAW (5)
2.	RC 9	1.500" (38.10)	BASE (1)		SHAFT SCREW (3)
3.	RC 9	0.400" (10.16)	BASE (1)		PLUNGER (9)
4.	RC 9	0.250" (6.35)	BLOCK (4)		SHAFT SCREW (3)
5.	RC 9*		PIN (7)		BLOCK (4)

*THE LIMITS OF CLEARANCE FOR THE HOLE IN THE BLOCK SHOULD BE ADDED TO THE FN 1 LIMITS OF THE PIN.

6. THE LIMITS OF TOLERANCE ON THE MAING DIMENSIONS OF THE VERNIER AND THE LUGS OF THE BASE ARE +0.00 AND -0.01 OR (+0.0 and -0.3 mm). THE BASIC DIMENSION OF THE VERNIER IS 0.75 (19.1 mm). THE ALLOWANCE BETWEEN THE TWO ASSEMBLED PARTS IS 0.01 (0.25 mm).
7. THE LIMITS OF TOLERANCE OF THE MATING DIMENSIONS OF THE BLOCK AND JAW ARE +0.00 AND -0.01 OR (+0.0 AND -0.3 mm). THE BASIC DIMENSION OF THE BLOCK IS 0.49 (12.44 mm). THE ALLOWANCE BETWEEN THE TWO ASSEMBLED PARTS IS 0.01 (0.25 mm).

PART 6: #6-32UNC-2A(M3.5 X 0.6) X .50(23mm) SOC HD MACHINE SCREW, MATL: STEEL

PART 8: .12(3mm) X .50(13mm) SPRING PIN, MATL: STEEL

PART 11 7/16-14UNC-2A(M12 X 1.75) X .37(9mm) HEX SOC FLAT POINT SET SCREW, MATL: STEEL

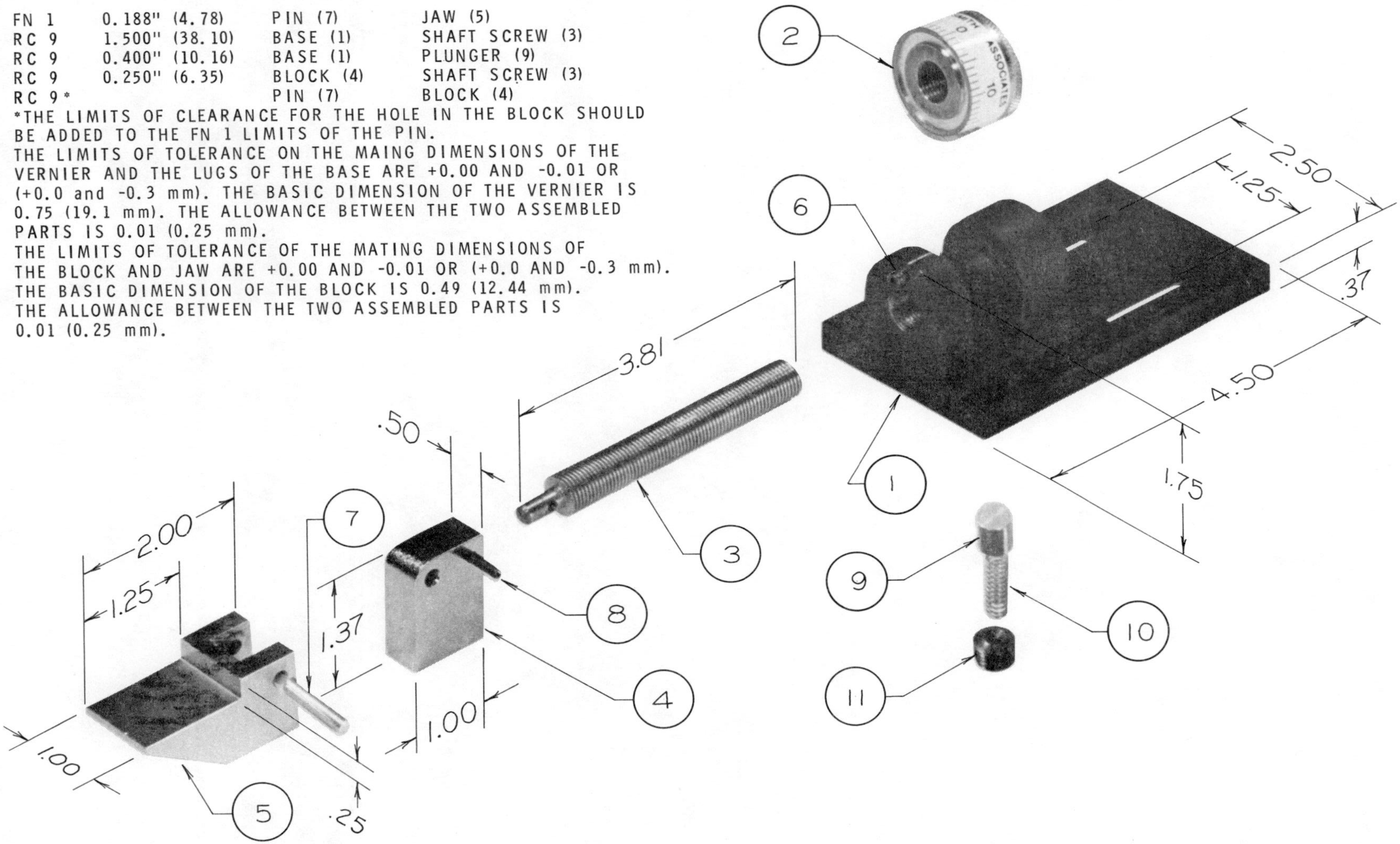

2.50 1.25 .37 4.50 1.75 3.81 .50 2.00 1.25 1.37 1.00 1.00 .25

1 2 3 4 5 6 7 8 9 10 11

PMM
PMM

WORK SHEET

Graphics & © Geometry	NAME:	MIN.	GRADE	
	FILE: SEC: DATE:			

WORK SHEET

Graphics & © Geometry	NAME: FILE: SEC: DATE:	MIN.	GRADE	

WORK SHEET

Graphics & Geometry ©	NAME		MIN.	GRADE	
	FILE	SEC DATE			

Graphics & Geometry	NAME	MIN.	GRADE	
©	FILE SEC DATE			

WORK SHEET

Graphics & Geometry	NAME:			MIN.	GRADE	
	FILE:	SEC:	DATE:			

WORK SHEET

Graphics & © Geometry	NAME FILE SEC DATE	MIN.	GRADE	

WORK SHEET

Graphics & ©Geometry	NAME FILE SEC DATE	MIN.	GRADE	

WORK SHEET

Graphics & © Geometry	NAME: FILE: SEC: DATE:	MIN.	GRADE	

WORK SHEET
Graphics & Geometry
NAME
FILE
SEC
DATE
MIN.
GRADE

WORK SHEET

Graphics & Geometry ©	NAME FILE SEC DATE	MIN.	GRADE	

WORK SHEET

Graphics & Geometry ©	NAME FILE SEC DATE	MIN.	GRADE	

WORK SHEET
Graphics & Geometry
NAME
FILE
SEC
DATE
MIN.
GRADE

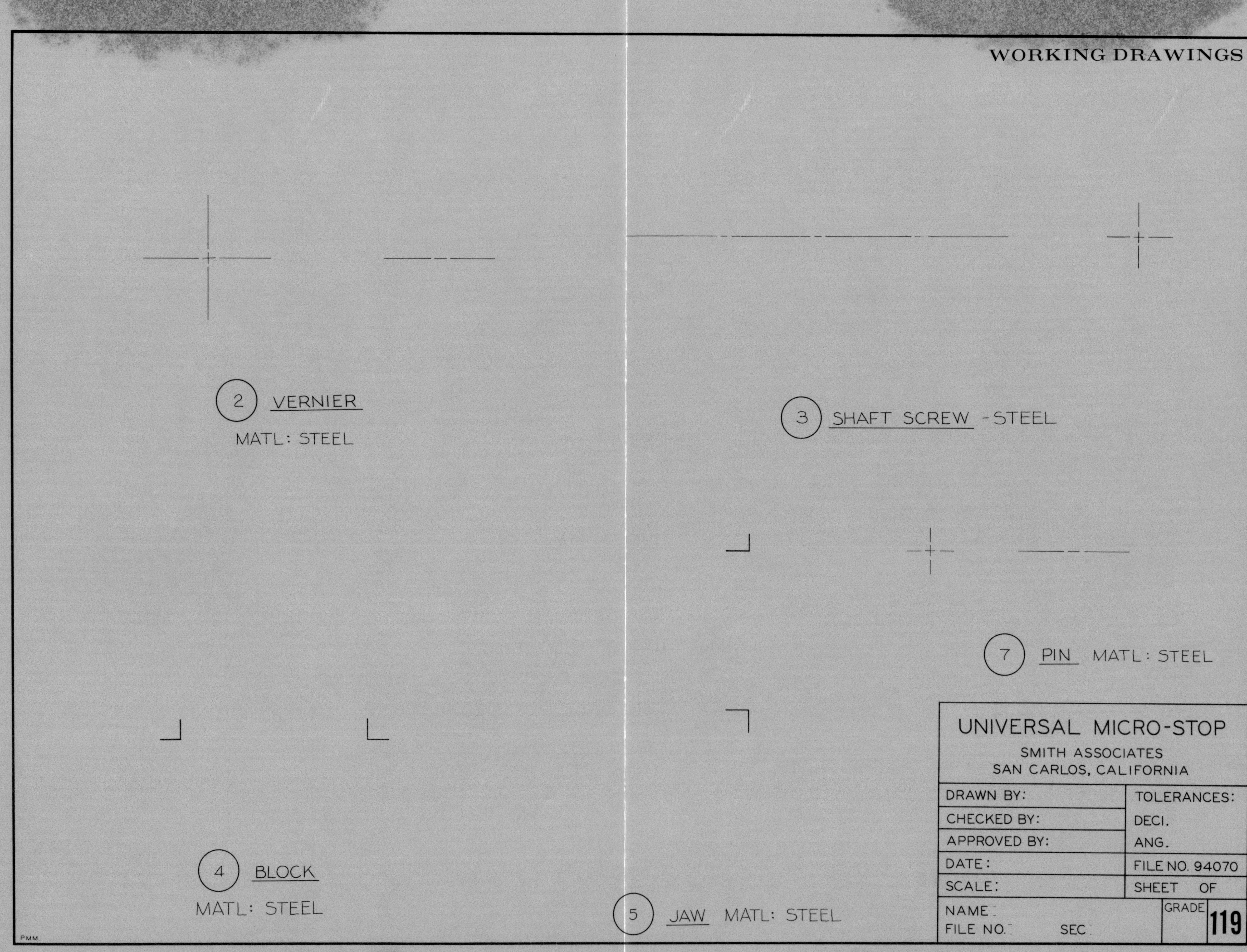

WORKING DRAWINGS
2 VERNIER
MATL: STEEL
3 SHAFT SCREW - STEEL
7 PIN MATL: STEEL
4 BLOCK
MATL: STEEL
5 JAW MATL: STEEL
UNIVERSAL MICRO-STOP
SMITH ASSOCIATES
SAN CARLOS, CALIFORNIA
DRAWN BY:
CHECKED BY:
APPROVED BY:
DATE:
SCALE:
TOLERANCES:
DECI.
ANG.
FILE NO. 94070
SHEET OF
NAME
FILE NO.
SEC
GRADE
119
PMM

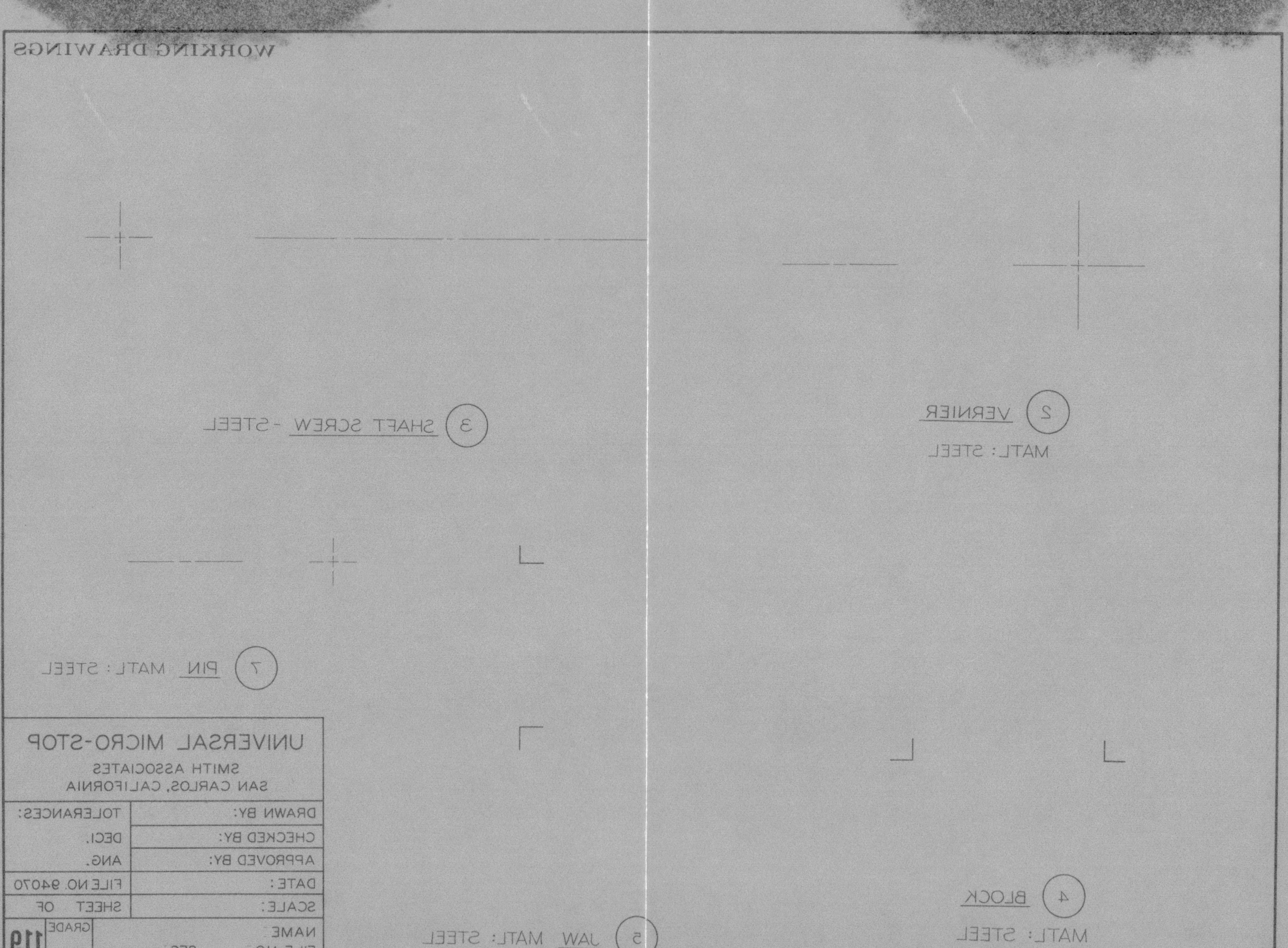
WORKING DRAWINGS
2 VERNIER
MATL: STEEL
3 SHAFT SCREW - STEEL
7 PIN MATL: STEEL
4 BLOCK
MATL: STEEL
5 JAW MATL: STEEL
UNIVERSAL MICRO-STOP
SMITH ASSOCIATES
SAN CARLOS, CALIFORNIA
DRAWN BY:
TOLERANCES:
CHECKED BY:
DECI.
APPROVED BY:
ANG.
DATE:
FILE NO. 94070
SCALE:
SHEET OF
NAME
FILE NO. SEC.
GRADE
119